FUNDAMENTAL STATISTICAL PRINCIPLES FOR THE NEUROBIOLOGIST

FUNDAMENTAL STATISTICAL PRINCIPLES FOR THE NEUROBIOLOGIST

A SURVIVAL GUIDE

STEPHEN W. SCHEFF
University of Kentucky Sanders-Brown Center on Aging, Lexington, KY, USA

ELSEVIER

AMSTERDAM • BOSTON • HEIDELBERG • LONDON
NEW YORK • OXFORD • PARIS • SAN DIEGO
SAN FRANCISCO • SINGAPORE • SYDNEY • TOKYO
Academic Press is an imprint of Elsevier

Academic Press is an imprint of Elsevier
125 London Wall, London EC2Y 5AS, UK
525 B Street, Suite 1800, San Diego, CA 92101-4495, USA
50 Hampshire Street, 5th Floor, Cambridge, MA 02139, USA
The Boulevard, Langford Lane, Kidlington, Oxford OX5 1GB, UK

Notices
Knowledge and best practice in this field are constantly changing. As new research and experience broaden our understanding, changes in research methods, professional practices, or medical treatment may become necessary.

Practitioners and researchers must always rely on their own experience and knowledge in evaluating and using any information, methods, compounds, or experiments described herein. In using such information or methods they should be mindful of their own safety and the safety of others, including parties for whom they have a professional responsibility.

To the fullest extent of the law, neither the Publisher nor the authors, contributors, or editors, assume any liability for any injury and/or damage to persons or property as a matter of products liability, negligence or otherwise, or from any use or operation of any methods, products, instructions, or ideas contained in the material herein.

British Library Cataloguing-in-Publication Data
A catalogue record for this book is available from the British Library

Library of Congress Cataloging-in-Publication Data
A catalog record for this book is available from the Library of Congress

ISBN: 978-0-12-804753-8

For information on all Academic Press publications
visit our website at https://www.elsevier.com/

  Working together
to grow libraries in
developing countries

www.elsevier.com • www.bookaid.org

Typeset by TNQ Books and Journals
www.tnq.co.in

Dedication

To my wife Sue, who suffered through my constant rambling about this work.

To my neurobiologist daughter Nicole, who always had another great statistical question.

To the many students I've had in the laboratory and in class who provided me with the incentive to write this book.

To my mentor Dennis C. Wright, who cultivated my love for science and experimental design.

Contents

4. Graphing Data

5. Correlation and Regression

6. One-Way Analysis of Variance

7. Two-Way Analysis of Variance

8. Nonparametric Statistics

9. Outliers and Missing Data

10. Statistic Extras

Index   209

Preface

Do you know what a Family Wise Error Term is? How about when to use the standard deviation and when to use the standard error of the mean? When someone says they have 85% power does that mean they don't need to plug in their digital device for a while? How about the difference between a Dunn's test and a Dunnett's test or the Wilcoxon two-sample test and the Mann–Whitney U statistic? If you don't then this book is for you. All of these terms are found in the neuroscience literature and many of them are used incorrectly.

Let me state up front—I'm not a statistician but I am a renaissance neurobiologist. I have dabbled in many different disciplines within the field of neuroscience and thus have some appreciation for the issues the researcher might encounter. I wrote this book to help graduate students in the field of neuroscience. The student who masters this material should be able to evaluate the statistical results of most papers in the field of neurobiology. A book of this limited size really can't begin to cover the field of statistics and that's not the intent of this work. I like to use the analogy of describing the extrapyramidal system. Where do you begin and what structures do you include? Is it important to mention the rubro-spinal tract and the tectospinal tract? How much detail do you go into when talking about the circuitry of the basal ganglia and the nigrostriatal pathway? The purpose of this book is really just to give the neurobiologist fundamental statistical principles, hence the name—a survival guide.

This is not a traditional statistics book—it contains almost no statistical equations or statistical symbols. It only covers some of the most basic situations that the neurobiologist will encounter. It is a book that hopefully will help develop and increase an understanding of statistical theory and practices. Virtually every researcher today has a personal computer that can be loaded with really fantastic statistical software or has access to one in the laboratory. Gone are the days when you had to memorize any statistical equations. I love working with these commercial programs and viewing the various outputs, even though some of these may appear to be uninterpretable. What I quickly realized was that these programs make very few assumptions about the data entered. In fact, they will attempt to process data without regard to its type or origin—which is a problem. These programs will provide the user with a multitude of different choices and simply wait for the researcher to make the choice. Some even provide interesting suggestions. But then that's part of the problem because most neuroscientists don't know which statistical routine is the most appropriate for a given set of experiments. Should you use the Tukey–Kramer or the Games-Howell multiple comparison procedure? And, of course, the programs can't really interpret the results. Part of the aim of this book is to provide the reader with information to critically choose which statistical test to command the software to use. I've tried to use very plain English and avoid jargon. However, one needs to also understand the language of the biostatistician and thus I've included some of that jargon with an explanation.

Although statistical reasoning/experimental design is crucial to experimentation it often never receives the classroom time it should. Most graduate students learn their statistics from other graduate students or from reading journal articles. Unfortunately many of the published articles, even in top-notch journals, report statistical findings that are either incomplete or incorrect. While many of the top biomedical journals have formal statistical reviewers that flag many of the most basic statistical faux pas, there are simply too many new and lower impact journals that don't scrutinize the data analysis components. Coupled with this is the overall lack of sophistication of the reviewers themselves who are often somewhat clueless of the difference between significance and effect size.

I have never envisioned that this would be the only statistics book that you will need. It is intended to be just a basic survival guide. There are so many great books on statistics and I often recommend several of these such as "Intuitive Biostatistics" by Harvey Motulsky and "Primer of Biostatistics" by Stanton Glantz. For the neurobiologist who wants a more detailed look at statistical theory and practice then I recommend the terrific text "Research Design and Statistical Analysis" by Myers, Well, and Lorch.

Many students are simply "afraid" to ask for help because they would appear as "less enlightened." They shouldn't be, because none of us are experts in the field of statistics. I still ask for advice when I dream up certain experiments. Your friendly neighborhood biostatistician is indeed your friend in science and wants to help you do your best. Although the student may not believe it, the biostatistician is trained to help you. That individual can assist you with both the design of an experiment and the data analysis. This text will help you make the most of your interactions with the biostatistician.

All of the examples that I use in this text are made up by me and as far as I know they don't exist in the literature. None of the experimental results have any foundation in reality. Actually, making up the data was one of the hardest parts of writing this book because I was always manipulating the numbers just to demonstrate a point and to focus on an important idea. I have created all of the graphs and tried to make them simple. Some of my own biases are shown in these graphs and represent my dislike for so many hard to understand graphs I've had the displeasure to navigate when reading grants for the NIH.

Although it isn't necessary to read the chapters in order, the first three simply present some of the basic concepts that every statistician "assumes" you know. I encourage the reader to carefully read the section on Type I and Type II errors because this is at the very heart of statistical reasoning. In Chapter 6 I tried to make sense of the field of multiple comparisons. This is a very confusing and heavily debated part of statistics but one that is crucial to the interpretation of data. Chapter 8 deals with nonparametric statistics, which are really very important in the field of neuroscience. You may have to read a particular section more than once to have a complete understanding of a point I'm trying to make. That's OK and expected. I've tried to make explanations extremely simple and easy to understand. The book covers a very wide range of statistical topics. Hopefully they will enable the reader to apply some statistical procedures properly so that the data can accurately reflect their meaning.

If after reading this book you have gained a new appreciation for statistics and have a greater insight into which statistical procedure is most appropriate for your data, then I have succeeded in my aim.

Steve Scheff
September 22, 2015
Lexington, KY, USA

About the Author

Steve Scheff is a Professor of Anatomy and Neurobiology at the University of Kentucky, with a B.A. in psychology/biology from Washington University in St. Louis. He earned an M.A. and a Ph.D. in physiological psychology from the University of Missouri in Columbia and spent 6 years as a postdoctoral fellow/staff scientist at the University of California–Irvine in the department of Psychobiology. He has been a member of the Society for Neuroscience since 1974 and served on way too many NIH study sections. He has published more than 200 scientific articles in the field of neurobiology. The author has taught human brain anatomy in the College of Medicine for more than 35 years and trained numerous graduate students and postdoctoral fellows the art of experimental design and statistics. When not playing scientist he chases a little white ball around the golf course, shoots competitive 8 ball and 9 ball, and tortures himself playing classical guitar.

"Rather than torturing your data until it confesses, treat it humanely and you will probably get an opera."

Anonymous

CHAPTER

1

Elements of Experimentation

Stephen W. Scheff

University of Kentucky Sanders-Brown Center on Aging, Lexington, KY, USA

Conducting data analysis is like drinking a fine wine. It is important to swirl and sniff the wine, to unpack the complex bouquet and to appreciate the experience. Gulping the wine doesn't work. **Daniel B. Wright**.

Any type of research begins with the formulation of a research hypothesis, which is usually derived from some prior information that has been used to create a theory about something. For example, changes in hippocampal proteins are the underlying basis for short-term storage of new information. This may or may not be true, but it is a theory about learning and memory that has been formulated because of past information. A scientist then formulates a research hypothesis that could be used to test that theory. The actual experimentation is then designed to test a particular hypothesis about the theory. Depending on the "quality" of the experimental design, one subsequently says that the data either support or do not support the hypothesis and then decides to change his/her understanding of the theory. Trying to work out the details of any type of investigation is actually a lot more difficult than one first imagines. You have a lot of decisions to make before you can collect the first piece of data. For example, there are many similarities between designing an experiment and hosting a "simple" wine tasting. Both of these undertakings require a lot of decision making and careful planning.

REASON FOR INVESTIGATION

The Wine Tasting

Let us say that you and your friends want to do a wine tasting and really want to do it right, in an unbiased fashion. After all, you are a neuroscientist and everyone likes to think their gustatory sensibility is beyond repute. But, if you stop and actually think about an unbiased wine tasting, you will quickly realize how complicated it is. The first question to consider is the reason for the wine tasting. Consider the following:

1. To impress my colleagues
2. To impress my "boss" and hope for a raise
3. To actually get an idea of what people like
4. To just have fun
5. To educate fellow colleagues
6. Part of project requirement

You might want to think of it as an experiment with a specific question: What is the hypothesis?—That there will be a noticeable difference between the different wines that even the uneducated palate can detect. For right now that hypothesis will work.

The Learning Experiment

If you are going to design a learning and memory experiment, you would also have to do something very similar to a wine tasting. You first have to ask yourself what is the motivation behind the experiment (e.g., pilot study for a grant submission or maybe even for a dissertation). The scope of the project will change accordingly. You need to state clearly the hypothesis—giving animals a neuroactive novel compound will alter their ability to learn a maze (see **Hypothesis** in Chapter 2).

WHAT TO TEST

The Wine Tasting

Are you going to do a red, white, blush, or sparkling wine? If you choose a red wine, will it be a cabernet sauvignon, merlot, zinfandel, petite sirah, pinot noir, charbono? If you choose a white wine instead, it could be a chardonnay, gewürztraminer, pinot gris, sauvignon blanc, chenin blanc, or a Riesling. Whatever type of wine one decides to evaluate, comparing a red to a white or a chardonnay to a Riesling would be like comparing a mouse to a frog. While they are both animals and vertebrates, they are entirely different. For the sake of simplicity we will try a cabernet sauvignon tasting. The next decision is which wines to purchase. This is certainly not a simple choice because there are many factors to consider. Again for the sake of simplicity you decide on domestic wines and will only evaluate those from northern California, although those from New York, Oregon, or Kentucky could have been chosen. Already there have been several very important decisions made and this will certainly impact your conclusions.

One still must decide on what vintage. This of course is also not a trivial matter, because if the wine is too young, it may not be ready for consumption and the tannin may be too high. If one chooses an "older" wine, then the availability is greatly reduced and the price per bottle can become a concern. So this of course goes back to the motivation for having the tasting. These are all very important variables. You simply decide to go for availability, drinkability, and price and thus choose a vintage within the past five years. Key here is that all of the chosen wines need to be the same vintage to control for possible differences in growing seasons. This is called a horizontal wine tasting, i.e., same type of wine, same vintage, just different vineyards. An alternative of course might be to do a vertical wine tasting, where the same wine and vineyard are selected, and several different vintages are evaluated (e.g., 1984, 1985, 1986, 1987).

The Learning Experiment

What novel compound are you going to investigate and what is the rationale behind it? Suppose you have heard that the novel compound

Smartamine® has a unique ability to recruit specific portions of the memory circuitry and may be involved in recovery from brain injury. Are there other very similar compounds that may also have some but not all of the same properties as Smartamine®? Are there different formulations of this novel compound?

LEVELS AND OUTCOME MEASURES

The Wine Tasting

If you only look at two wines it becomes a little boring so maybe three or four or five. The more wines that are decided on the more difficult the ranking. Just how will these wines be rated? Will you be using the Wine Spectator 100 point scale (which really only goes from 50 to 100) or the Robert Parker/Wine Advocate scale? Perhaps you just want to use a 20 point scale, or if you have just five wines a 5 point scale. Is this important? Actually it is very important and there are a number of considerations that need to be made. If you use the 50 to 100 scale, what is the individual's basis for a 100 point wine? Has anyone in your intended group actually tasted a 100 point wine and can they actually recall what that was like? What it then comes down to is determining what type of data you want. As you will see later, there are four different types of values (nominal, ordinal, interval, and ratio), and only certain types of statistical analysis can be carried out on the different types of values. It might possibly be the case that you and your guests really like all of the wines and only one really "stands out."

Perhaps you need to provide your guests with a little lesson on how to evaluate a red wine so that everyone at least attempts to use the same criterion. A classic faux pas is to tell someone to rate something from a scale of 1 to 10 and never indicate what a 1 or 10 actually signifies. Is a 10 the best or the worst? Is it possible to have a 6.5 or should only whole integers be used? If five wines are simply ranked on a scale of 1–5 and the rank of 1 is the best, and two of the wines cannot be differentiated from each other, do they both get the same rank or the average of the two ranks (e.g., both get a 3 or both get a 3.5)?

The Learning Experiment

In your animal learning experiment you will have to decide what concentration of the novel compound to use and whether or not the compound should be injected or taken orally. This of course will depend on the animal species on which you decide to test your compound. You probably will want to try several different amounts of the compound and explore a possible dose–response curve. Let us suppose that you do decide to use rats to test Smartamine®. You have come up with another significant decision as to what strain of rat to use. This of course will be contingent on the task that is chosen to study learning. If you decide to use a visual task then a rat strain with pigmented eyes would be preferable. Many times individuals choose to use Sprague–Dawley or F344 rats for the Morris water maze task and require the subject to use extramaze cues to locate the submerged platform. The visual acuity of an albino rat is quite bad and hence the investigator may unknowingly be making the task harder than it should be. The investigator will also have to decide what age the rat should be at the start of the experiment. This factor is not trivial since depending on the task, older animals move much slower than younger animals and very young animals may respond to the compound very differently. Finally there is the matter of gender in these experiments. It is very rare that both male and female rats are used in learning experiments and they may respond very differently to the novel compound.

What dependent variables will be tabulated as a quantitative measure of learning? Will it be errors to criterion or percent of correct responses, escape latency or path length? How many different trials will be given on a single day and for how many days? It is critical that one carefully projects what type of data will be collected and envision how that data might be presented graphically.

SITE PREPARATION AND CONTROLS

The Wine Tasting

For the present example you have chosen to do a horizontal tasting of 5-year-old cabernet sauvignon wines from California. The price range will be $15.00 to $19.99 so it will be a tasting of a domestic cabernet sauvignon for under $20.00. In this regard the range of wines has once again been "controlled" to some extent and it is very important to "attempt" to have some control over as many different variables as possible. You might ask your guests to each bring a bottle but then you run the risk of having some duplicates. In some scenarios this might be a very legitimate consideration.

How the wine is presented is also extremely critical. There is such a factor as name bias. I always like to talk about the difference between Doña Paula Los Cardos cabernet versus the Gnarly Head cabernet. Immediately, without even trying either one of these wines, tasters assign a better score to the Doña Paula Los Cardos cabernet, simply because the vineyard sounds more sophisticated than the Gnarly Head vineyard, when in fact both wines may be equally ranked. Usually one tries to "blind" the tasters so that they do not have this problem. Simply using a brown paper bag or covering the label could possibly be a problem. One of the wines may have a screw cap instead of a cork, which some individuals believe is the sign of a lesser wine, again introducing bias in the evaluation. One solution would be to decant all the wines into identical vessels, for example, identical laboratory flasks. Each flask is subsequently given an identifier such as a letter of the alphabet or a number like 872.

There is considerable literature about decanting a wine and letting it "breathe." Even though almost all of the less expensive wines are routinely filtered to a crystal clear state, it is possible that some may have some solid matter if they have not been stored correctly. Of course one would not want to have his/her guests tasting unsightly looking wines because should not the visual experience be part of the rating? Many wine critics claim that decanting the wine also brings those "young" wines into contact with the air and this markedly alters the wine. Young wines begin to "soften" and decanting helps to down play the tannic structure. Of course this will allow the wines to be brought to room temperature as "every great cabernet should be for tasting." One will simply have to take for granted that the wines chosen were properly stored prior to your purchase, although some may have been sitting on a loading dock in the sun at 97 °F for a day or two.

The Learning Experiment

Where will the actual testing take place? Is the room very close to the vivarium where the animals are housed? If the animals have to be transferred from one room to another, how much time is required for the subjects to acclimate to the new environment? Animals that are agitated and stressed can have a totally different response to not only a novel compound but also to the environment. Depending on the task, the actual room environment is extremely important with the appropriate lighting and ambient temperature. This can often be controlled by first handling animals several days prior to the actual testing in the testing environment. With rodents, the sense of smell is extremely important and the ability to let the subjects habituate can be instrumental in a successful experiment. External noise can also be an important factor and in some situations it is necessary to have a certain amount of "white" noise to negate possible other auditory cues, essentially forcing the rats to focus on a visual or perhaps a tactile cue. Equipment and space availability are also extremely important and overlooked. Will the facilities be available on a regular basis for the entire duration of the experiment? All subjects need to be evaluated under the same conditions. If one is using a water-related task such as the Morris water maze, the temperature of the water is extremely important and needs to be held constant between learning sessions.

How will subjects from the different groups be coded so that the experimenter(s) is(are) blinded to group designation. This is perhaps one of the greatest mistakes that can be made during the actual execution of an experiment. All data must be collected before the grouping code is revealed. If a daily injection of the novel compound is part of the experimental protocol, and the principal investigator is also the individual administering the drug and testing the subjects, it is imperative that another individual randomly assign the subjects to specific groups. The use of a vehicle or placebo compound is mandatory. Identical containers need to be used for both placebo and experimental drug with totally generic labels.

TROUBLESOME VARIABLES

The Wine Tasting

Will the wine be tasted in appropriate glass stemware, plain small glasses, plastic stemware, plastic glasses, paper cups, Styrofoam cups, ceramic coffee mugs? One can only imagine that this could also have a significant bias as to the rating. If visual appearance (color, clarity) is part of the rating the tasting vessel needs to be transparent. Room lighting needs to be sufficient. Will the same glass be used for each of the wines or will separate glasses be used so that an individual can more closely compare two wines in terms of bouquet, color, body?

If a tasting was held at 6 am as opposed to 6 pm, for many people this would most likely bias their evaluation of the wines, unless of course they all work the "grave yard shift" and 6 am is their normal "party time." Perhaps the tasting is going to take place immediately after a seminar dealing with "The Anatomy of Taste Receptors" or "Orexin–Corticotrophin Releasing Factor Receptor Heteromers." The actual environment for the tasting can impose a significant bias as to the evaluation process. Consider the possible difference of having a tasting in your own home or at a complex clubhouse versus holding it at the chairman's or institute director's house. Is the tasting going to last 1, 3, or 6 h? Will everyone be able to sample each wine only once or multiple times and will there be a specific order assigned to the tasting? Perhaps a designated "wine pourer" will fill each glass with a specific amount. What is that amount? If each individual has a 6 oz glass, those wines sampled first will have a different score than those sampled later. Also, how is the palate "cleansed" between wines? Is some type of water available or perhaps some type of "cracker"? It is possible to become dehydrated drinking wine. Of course, wheat crackers are on the forbidden list because some guests may have a gluten allergy. Perhaps iceberg lettuce would suffice.

One extremely important variable for any wine tasting is what the individual ate prior to sampling. Consider the possibility that the night before some of the guests were at a Mexican chili cookout with an accompanying tequila tasting. Or some of the guests decided to try the new Thai restaurant immediately before the tasting. Perhaps this could be controlled by first having everyone involved in a light meal prior to the tasting.

The Learning Experiment

For many learning experiments the animals need to be tested at the same time each day. If different groups are used because of the necessity to use increased numbers then time of day is an important variable. It must be ensured that the animals are maintained on a constant light/ dark cycle. Unless the experiment has something to do with circadian rhythms or stress, the environment for each of the subjects needs to be as closely controlled as possible. That could easily mean that something as simple as cage maintenance by the vivarium staff immediately before testing could have a strong influence on learning outcome. This can be partially controlled by working with the vivarium staff and noting when cage maintenance is scheduled or if it can be rescheduled. During the actual testing what happens to the animals between trials? Do they remain in separate cages or in group cages? Is the area a well-lit or a dimly lit environment? Is testing to be carried out by the same investigator or will a "team" of investigators be involved in handling the animals? The familiarity of the investigator involved in the testing and their animal husbandry skills can totally bias a learning situation. Just the simple way an animal is removed from a holding cage and placed in the testing environment can totally bias a particular trial or set of trials. If multiple testers will be involved they must all perform the task in the same fashion.

Statistical Analysis of the Data

Unless the data are gathered appropriately and analyzed using the correct statistical tests, the interpretation of the data will be problematic. What type of data will actually be collected? What happens if a particular guest decides not to taste all of the wines? What happens if one of rats gets sick and cannot continue with the learning experiment? All of these scenarios have to be decided on before any data are collected.

WHAT DO YOU DO FIRST WHEN YOU WANT TO RUN AN EXPERIMENT

Remember: KISS (Keep It Simple and Scientific)

First: Try and clearly state what it is that you are trying to investigate and see if you can come up with a hypothesis (Chapter 2). The key to good experimentation is asking the right question and in the preparation.

Example: *Does the novel compound Smartamine® make you smarter?* While this is certainly a "scientific" question it requires considerable refining. For instance, one has to define what variable will be used to define "smarter." Then, of course, there is the question of how much of the compound will be administered and when will it be administered prior to testing? Will the compound be administered over an extended period of time? Who will the subjects be? These are just a few of the many variables that need to be defined.

Refined question: Does a diet containing 2% of the novel compound Smartamine®, when fed to young adult male Wistar rats over 30 days, reduce the number of errors to criterion in the radial-arm maze.

Second: Go to the literature and read as much as possible about the topic you are interested in studying (e.g., Smartamine® and the radial-arm maze). Try to figure out if there is a workable approach to test your hypothesis. For the above hypothesis you might question whether a 2% laboratory diet would be feasible and how it would be administered.

Third: From the literature try and decide how much variance you might expect to see for the dependent variable you are going to study. For example, what range of values would a typical rat demonstrate in the radial-arm maze? About how many trials and errors to criterion are average for the radial-arm maze?

Fourth: Make a decision on how robust an effect you would consider a significant reduction in number of errors to criterion (e.g., 30% difference between groups or a 15% difference).

Fifth: Try and list as many different important variables that need to be controlled. To name a few: how are the subjects housed (single or group cages); what time of day will the testing take place; where will the testing take place and what are the environmental conditions; how is the maze cleaned between trials; are the animals given ad libitum access to the food prior to testing; what is the location of the maze in relation to the animal housing; who will be doing the testing; will the experimenter be in the room with the maze; will the animals be tested for locomotor ability prior to testing; what is the control diet; does the Smartamine® make the feed look or "taste" differently; does the experimental diet make the animals more active or does it affect important sensory cues (e.g., more sensitive to light). **Sixth**: Make an appointment with a biostatistician and discuss your ideas and research plan. Your biostatistician may want to know your rationale for your investigation. For the above hypothesis, the statistician will definitely want to know how you are defining the "learning criterion," if the task has been used in the past, and if so what kind of data other studies have shown. He/she will want to know what effect size you think is important and about how many animals you may want to use. Finally he/she will want to know what your tolerance is for possibly making a type II error (Chapter 3). **Seventh**: Decide what statistical approach you want to use to analyze the data. The biostatistician will help with this.

TYPES OF EXPERIMENTAL DESIGN

Often when an experimenter approaches a biostatistician he/she is asked "What is the basic experimental design?" This might seem like an unusual question but it is extremely important in determining the type of statistics that are appropriate. There are basically two different types:

1. Observational: Researcher looks for existing differences in populations of subjects
2. Manipulational: Researcher alters some condition and looks for differences

Manipulational experiments are often regarded as internally valid experiments. That means that the research design controls for all possible alternative explanations other than the one being investigated. These types of experiments actually require control over multiple variables and usually require a laboratory setting. The experimenter manipulates one or more factors and quantifies the outcome of one or more dependent variables. There are basically two different types of manipulational experimental designs: **(1) between-subject design; (2) within-subject design**.

Between-Subject Design

The most basic type of experiment is testing possible differences in two independent groups of subjects that have had some type of different treatment, e.g., mice given an **experimental drug** versus mice given the **vehicle** and subsequently tested for changes in maze performance. Essential to this type of design is the total **random assignment** of subjects to either the drug or vehicle treatment group. Every subject in a particular group has been **treated exactly the same**. For example, if the design of the experiment is to test how the novel compound alters maze performance in rats, then each rat in the drug group needs to receive the drug at the same time prior to testing and at the same dose/body weight, using the same vehicle. The vehicle-treated animals also need to be treated at the same time prior to testing and at the same vehicle volume/body weight. Since the animals assigned to each group are randomly assigned, this negates possible biological differences, and one has a **completely randomized design**. The absolute key to these experiments is the lack of bias in the group assignments. If this is done properly, then there will be an equal number of subjects in each group. Even if there are multiple different doses of the experimental drug being tested, the assignment must be random.

Let us say one has 10 mice in each group, and there is only one manipulation of the drug and only one measure of maze performance (e.g., time to complete maze)—there would be a total of 20 data points, one data point for each mouse. Most experimental designs are planned with the same number of subjects in each group. This provides equal weight in the subsequent statistical analysis. However, sometimes there is a loss of subjects in a specific group. For instance, in animal experiments there is sometimes a loss of data or the fact that a particular subject, following some type of surgical manipulation, cannot perform the task. If the loss occurs at the very start of the experiment, before any of the animals have undergone novel treatment that is not common to all groups, then simple replacement can occur. If the loss occurs later on, such as after drug treatment, then it is not appropriate to simply add additional subjects, **because the experimenter has lost the randomness of the assignment**.

In a recent set of experiments, an investigator was testing whether or not a novel compound, when given post trauma, could enhance learning of a spatial memory task. Ten animals were designed to be in either the drug or the vehicle group. During the surgical manipulation of inducing the trauma, one of the animals expired. Since none of the animals had been assigned to either the drug or the treatment group, the experimenter could replace the animal with another as long as it was the same age, gender, etc. After the surgical manipulation the animals were assigned to either drug or vehicle. If the loss occurs because of an experimental procedure that is not common to all groups then simple replacement cannot be done because of the loss of randomness. Before running an experiment, it is important to carefully evaluate the protocols and determine whether or not the manipulations in the experimental group could inflict more stress on these subjects compared to the control group. This is the reason why animals in the control group receive injections of vehicle with the same volume and route as those in the experimental group. When subjects or data are lost during the course of an experiment, additional subjects cannot be simply added to make the groups equal, even if the subjects are drawn from the same population. Many statisticians simply recommend that at the end of the experiment, subject data from the other groups be randomly chosen and removed to make the groups equal. Alternatively, there are statistical methods that can be used to deal with unequal subject numbers per group as long as they are not substantially different (see Chapter 10).

When the loss of subjects is substantial and it is necessary to increase the number of subjects to obtain sufficient power, then an equal number of subjects needs to be added for each group. This is often ignored by many investigators because of the cost of the experimentation. It is not uncommon to learn of investigators adding many additional experimental subjects and only one or two control subjects to make the total number of subjects equal per group. It is also <u>inappropriate to simply increase the number of animals in the experimental group to compensate for the potential loss later on</u>. When an experimental condition results in a greater loss of subjects in a particular group, the remaining animals represent a biased population, because they are perhaps "healthier" or "stronger" and more capable of surviving the manipulation.

Within-Subject Design

The above experiment could also be carried out a little differently. Subjects could first be tested with the drug and then be tested without the drug. Or they could be tested without the drug and then with the drug. This type of experiment adds complexity to the data analysis and is considered to be a **repeated measure type of design**. It is repeated measure because data are collected from the *same subjects* under *different conditions*. In this type of design it is again essential that subjects be randomly assigned to either the drug or the vehicle group. Half of the subjects would first have the drug followed by testing and after some time treated with the vehicle and evaluated. The other half of the animals would first be evaluated with vehicle alone and after an identical time as that in the drug group treated with the drug and evaluated. The advantage of this type of design is that it again controls for possible biological variance. This type of design is also called a **crossover design**. There are special statistical methods used to evaluate data for a repeated measure design that will be discussed in later chapters.

Why is it necessary to first treat some animals with drug or vehicle and then reverse the order? Why not just treat all the animals first with the vehicle and then with the drug? In some experiments this could be justified such as electrophysiological recording from slices in a dish. One evaluates outcome under a "physiological" normal condition and then under some manipulation. In the case of maze performance, there could easily be an effect of familiarization with the maze that could bias the outcome.

There are many books written entirely about experimental design that are really excellent. The curious student of statistics may want to browse Kirk[1] or Myers et al.[2]

SUMMARY

- When designing an experiment, list all possible variables that need to be considered.
- Read the appropriate literature and make the research question as exacting as possible.
- Keep the design simple and scientific.
- Consider what type of data will be collected.
- Discuss the research design with a biostatistician.

References

1. Kirk RE. *Experimental Design*. London: SAGE Publications; 2013.
2. Myers JL, Well AD, Lorch RF. *Research Design and Statistical Analysis*. New York: Routledge; 2010.

2

Experimental Design and Hypothesis

Stephen W. Scheff

University of Kentucky Sanders-Brown Center on Aging, Lexington, KY, USA

Fancy statistical methods will not rescue garbage data. **R.J. Carroll (2001)**

OUTLINE

Fundamental Statistical Principles for the Neurobiologist
http://dx.doi.org/10.1016/B978-0-12-804753-8.00002-6

15

Critical to any type of experimentation is first, stating precisely the question that one is trying to answer, and secondly, carefully planning the experimental procedures that attempt to answer that question. While many neuroscientists may believe this is self-evident, the primary reason so many experiments do not contribute to our basic understanding of a particular problem is a failure to correctly carry out these two basic components. Before beginning any actual experimentation, it is extremely important to state what the hypothesis is that you are trying to investigate. The American statistician, Raymond Carroll, is credited with saying, *"Fancy statistical methods will not rescue garbage data."* The way you get garbage data is by not asking the right question, and by failing to design the right approach.

HYPOTHESIS—ASKING THE RIGHT RESEARCH QUESTION

Statistics cannot prove or disprove a certain claim; e.g., if you eat carrots every day you will improve your visual acuity. All statistics can do is support or not support a hypothesis: if a rat receives a cortical contusion it will have difficulty learning a Morris water maze. Before you can test an idea, you have to formulate a research hypothesis. A **hypothesis** is simply a statement that deals with the relationship between some variables that often results from some type of acquired data. For example, you believe that in order for a rat to learn a maze it requires cognitive ability. You also believe that if a rat has head trauma it will disrupt the brain and that should result in a change in cognitive ability. Why not hypothesize that following a cortical contusion there will be a deficit in the animal's ability to learn a maze?

But what is this hypothesis really asking? What this hypothesis really says: if I get a rat from the vivarium, and subject it to a cortical contusion, it will not do as well as a naïve rat, when tested in a water maze. Making an inference, about a general population of rats in the vivarium, is the actual motivation behind the hypothesis. What you are really planning on doing is selecting a sample of rats from the very large group in the vivarium and, after administering some type of procedure, making some type of inference about the entire rat population. The bottom line in a research hypothesis is making some type of assumption about a characteristic of a **population**.

Population versus Sample

The question becomes, what is a **population** and what is a **sample**? A population is a large, perhaps infinitely large, group of "individuals," "objects," or "anything" that has a specific type of characteristic. In our example, the population is all the rats that have a specific type of cortical contusion, and not limited to just those in the local vivarium. As a researcher, we want to make an inference that others can apply to their animals with the same type of injury. If we assess a subset of these rats it is a **sample**. Normally, the population of interest is large enough that it is not practical to assess all the individuals. The term individuals here actually can refer to students, cats, dogs, rats, trucks, laptop computers, jelly beans, etc. When formulating the research hypothesis, it is critical to define the population of interest. Neuroscientists usually only deal with samples, which are basically subsets of the population of interest. They assume that the sample is an accurate estimate of the population because the sample was drawn randomly.

In numerous published research papers you will see statements such as, "Young adult rats were used as subjects…" This is not a well-defined research population. A better statement would be, "Young adult male Wistar rats, 200–250 g, were used as subjects…" One now has a clarification of the rat strain and also some indication of what the term young adult actually means. In the experiment, since the investigator does not have the funding or the time to study all of the Wistar rats with brain injury, they will be testing a small subset of this population—**the sample**.

What if you have developed a unique transgenic mouse, and you alone have all 15 of these animals and test them for their ability to learn a water maze, are you evaluating a sample or a population? In this case you have a very small unique **population** of transgenic mice that you are testing. You do not have to make any inferences because you know exactly how the population responds. This is very different from taking a small group of rats, subjecting them to some type of manipulation, and claiming this was a population. This latter group is merely a sample of a manipulation applied to the larger population. However, if you are studying your very unique population, you will have a difficult time generalizing your results to other transgenic mice.

One Population or Two Different Populations?

When we evaluate two samples with appropriate statistical tests, we attempt to determine if the two groups are truly different based on some variable of interest. If in fact the test shows that the two samples are indeed different, then we conclude that they came from different populations. If the results of the statistical analysis say that they are not statistically different, then we state that they came from the same population. Because samples are not perfect representations of the population they were drawn from, one cannot ever be certain that the conclusions drawn from the samples are *absolutely* correct. There is always a chance that the effect we observe is due to chance, and this is what statistics is all about.

NULL HYPOTHESIS (H_O) AND ALTERNATIVE HYPOTHESIS (H_A)

Before you can begin to formulate a hypothesis, a somewhat detailed review of the literature is a requirement. This is perhaps the No. 1 piece of advice that I can give to any aspiring or seasoned researcher. When I advise someone to read the literature it must be more than perusing a few review articles on the topic. While this may sound like heresy, 95% of the review articles do not mention some of the *most important* studies in a specific subfield of neuroscience. The qualifier here is the words, "most important." The authors of a review treatise have some definite biases and select those studies that primarily support their objective. Very commonly, the interpretation of studies in the review article is biased because the authors have a preconceived idea of what the literature should say and mean. A good strategy is to first find a couple of reviews (if they exist) and then look up some of the very recent studies that are cited in the review. Actually, by reading some of the primary references, one can begin to make a judgment as to how the author of the review has been interpreting the literature and may in fact give actual "clues" on how a research design should or should not be constructed.

Here is an example of a typical research hypothesis:

The novel compound Smartamine® makes rats smarter.

While this is certainly a "scientific" question, it requires considerable refining. For instance, one must define what variable will be used to quantify "smarter." Then of course, there is the question of how much of the compound will be administered and when will the compound be given prior to testing. Will the compound be administered over an extended period of time? Who are the subjects to be tested? These are just a few of the many variables that need to be defined.

Refined research hypothesis: A diet containing 2% of the novel compound Smartamine®, when fed to young adult male Wistar rats (250–275 g) over a 30 day period, reduces the number of errors to criterion in the radial-arm maze compared to rats fed a standard lab diet.

This research question is actually pretty straightforward and easy to understand. It directly translates into a testable set of experiments. This refined research hypothesis makes a statement concerning the feeding of a novel compound and a cognitive test, two important variables. The above research hypothesis asserts a positive relationship between the novel compound Smartamine® and the learning, and in fact most research hypotheses state positive relationships. It also defines what is meant by the term smarter (number of errors to criterion in the radial-arm maze). However, from a statistical perspective, we are interested in a different hypothesis. The primary hypothesis to be tested statistically is called the *null hypothesis*, which is often abbreviated **H$_O$**. This null hypothesis simply states that giving animals the novel compound Smartamine® is not going to make them any less prone to making errors in the radial-arm maze. This is the direct opposite of what statisticians call the *alternative hypothesis* (**H$_A$**). The alternative hypothesis is the one that most researchers "think" they are trying to test and the one they tout in their research results. In this case, the H$_A$ states that Smartamine® is going to make the Wistar rats perform better in the maze.

Why do statisticians make such a big deal about the H$_O$ versus the H$_A$? The answer to this question is actually quite simple. *It is the null form of the research hypothesis that is actually being tested statistically.* This should probably be placed in neon lights. The way that a research hypothesis is stated makes a major impact on how the statistical outcome is interpreted. This is a very important point because when you talk to your friendly neighborhood biostatistician he/she is all about the null hypothesis.

When the null hypothesis is stated there are actually two judgments that can be made: **Decision No. 1**, *accept the null hypothesis*; or **Decision No. 2**, *reject the null hypothesis*. Decision No. 1 is reached if the statistical analysis fails to reach some predetermined level of difference between the groups. One would say that the groups are not statistically different. A common statistical tool used to test the difference between two independent groups is the Student *t*-test (Chapter 6). With this test you set a level that you think would indicate a realistic difference between groups, and this is called the alpha level. You then test to see if that level of difference between groups actually exists. What the null hypothesis says is: "this predetermined level of difference between the groups that you set does not exist." What does that actually mean? If you perform your statistical test (e.g., Student t-test, with alpha set at 0.05), and you look up the results of this t-test in the appropriate statistical table, and it does not equal or exceed a critical value in that statistical table, then you **accept** the null hypothesis. Your test would say, "There is no statistical difference between the groups for the predetermined level that you set." If that value does equal or exceed the value in the statistical table, you **reject** the null hypothesis and accept decision No. 2. You conclude that there is a difference between groups at the predetermined level that you set. In this example, why is this statistical table so important? For the Student t-test, the statistical table is called the **t-distribution**, and it is the **probability distribution** that corresponds to the statistical test you are using. Remember that all of statistics is based on probability. Fortunately, commercially available statistical software packages do all of this nasty dirty work for you, such as looking up values in the t-distribution. These software programs tell you what the statistical significance is for your comparison. However, **they cannot interpret the meaningfulness of the outcome**.

WHAT IS PROBABILITY ANYWAY?

In terms of statistical reasoning (frightening as it may sound) probability is the likelihood that something will happen by chance. It is the basis for inferential statistics and the "normal curve." You might say that you are 99% sure that the cream you put in your coffee was not expired but then…maybe you read the expiration date wrong and thus there is a 1% chance it is expired. There is a 50–50 chance that when you go home your pet hamster will be exercising in its running wheel. So 50% of the time it is using the wheel and 50% of the time it is not. Of course it will depend on what time of the day you go home. So there are really several different ways that one can state probability in statistics. If there is a 100% chance that something will occur then it can have what statisticians call a p value $= 1$. If there is a 50% chance then the p value is 0.5, and a 10% chance is 0.1, and a 1 in a 100 chance is 0.01. What this 0.01 means is that there is a 1% chance that the opposite will occur. There is a 1% chance that the null hypothesis is correct. A p value of 0.005 means a 1-in-500 chance that the null hypothesis is correct.

Let us say that you ordered 50 "young" adult rats from Rattail Farms for your learning experiment and weighed each rat when it came in. You expected all the rats to weigh between 255 and 270 g. But you find that five of the rats were only 240–250 g, so 10% of them were underweight. Does this mean that Rattail Farms does not have enough rats of the same size to fill your order? It may be that the worker filling your order simply grabbed some rats from the wrong cage and sent them to you. Or maybe they did not have any more in the weight range that you wanted but wanted to send you 50 animals. Or maybe every order that they send out has a 10% error in the weights of the rats. You really do not know unless you make another order. So you place another order and this time only two animals were not in the weight range you selected (both overweight this time). Each order that you make is actually a sample of the young adult rat population at Rattail Farms. It might be the case that a third order would not have a single underweight rat or it may have four underweight and seven overweight rats. So, what is the probability that you will get exactly what you ordered from Rattail Farms? However, in so many different cases, the experimentalist makes a preliminary judgment on the basis of the outcome of a single experiment from a small sample. Ideally, what one wants to do is repeat an experiment several different times and gain a better idea of what the "typical" results should be. However, this costs a lot of money and also a lot of time. So one must decide what the consequences are of repeating or not repeating an experiment. Statistical tools can certainly shed light on the probability of a particular event or outcome occurring or what might be typical, but these tools cannot actually tell you what to do. In the case of the rats from Rattail Farms, you might want to try purchasing from a different vendor such as The Rodent Ranch.

STATISTICAL SIGNIFICANCE

What do we mean by **statistical significance**? What this simply means is that the differences found when comparing two or more groups have a certain probability of not occurring by chance. Statisticians commonly talk about populations and it is important to understand what this actually means. In the above experiment with the novel compound Smartamine®, the researcher believes that there are two entirely different populations. One population is all the animals that could get Smartamine® (even though only a few may), and the other population is the population of animals that gets fed just the standard diet. The researcher performs the experiment hoping to demonstrate that there are indeed two different populations. The statistician, by stating the null hypothesis, is simply saying, "No, there is really only one population, and the manipulation you have introduced (giving some animals Smartamine®) does not do anything to alter the learning ability in this population."

So you run the test and there is a significant difference between the mean learning scores of two groups and you say, "Statistically speaking, the two groups were different at the 0.05 level."

And the statistician says, "Well, you do understand, there is still a chance (5 out of 100) that there is not a difference, and in fact all of these animals actually belong to the same population. What really happened is that you just got lucky and tested the right animals."

So who is right, you or the statistician? Both of you could be right so you would have to do more testing. By running that statistical test, you basically determined the probability that the difference between the groups was in fact a real difference. Consequently, based on this probability, you attempt to draw a conclusion about the "reasonableness" of the null hypothesis. Remember, it is really all about the null hypothesis, which can **never** (*and I do mean never*) be totally rejected or accepted with complete certainty. As an old statistician once told me, "The null hypothesis is either very improbable, extremely probable, or somewhere in between."

WHAT IS A SIGNIFICANT EXPERIMENT?

Almost every scientific experiment makes some type of assessment as to whether or not the outcome supports or does not support a specific hypothesis. If the experiment involved statistical analysis then the level of support is given in terms of a numerical probability value such as 0.05. What this value actually provides is a measure of how sure the experimenter can be about the results. In this case, the experimenter is saying that there are 5 chances or less in 100 that there truly is no difference between experimental groups. To state another way, there is a 5% chance that the difference in the groups was simply a fluke, or happened by an error in random sampling, or perhaps some unknown variable had a role to play. If you told a statistician that you obtained a significant result, they would ask what alpha level was used to determine significance, or what p value was used to test the null hypothesis.

Selecting a p Value

This is actually not a trivial matter and is really based on the circumstances and consequences of the outcome of the experimental design. The most common practice is to set the level of significance at 0.05 or below. But it may make sense in some cases to make it 0.1. The actual origin of this 0.05 level of significance is very interesting[1] and may have appeared in print in 1925 for the first time in a statistical text written by a very famous statistician, Ronald Fisher. What is quite clear is that the level of significance is really based on probability, as is the entire field of statistics. When running a pilot study using a very small number of subjects, it might be wise to use a p value of 0.1 or 0.15 since a single data point could easily "sway" the difference in either direction. One certainly does not want to accept the null hypothesis when in fact the alternative may indeed be true.

Type I and Type II Errors

Most scientists run an experiment on a single sample of a population and thus use this one chance of deciding whether or not to pursue a particular hypothesis. Often this is the "pilot" study or the experiment after a very limited pilot study. It is important then to draw the correct conclusion concerning support for either the null hypothesis or the alternative hypothesis.

There are basically <u>two primary types of error</u> that you can make regarding conclusions concerning your hypothesis. Naturally there are many different errors you can make actually running the experiment and designing the experiment and reporting the experiment, but in terms of the hypothesis there is a total of four different outcomes and two of them are errors. Let us take as an example testing whether or not the novel compound Smartamine® makes rats learn a maze faster than rats given a placebo. The <u>actual truth is</u>, if all the young adult male rats at the breeders were properly tested, it would show that <u>Smartamine® **does not** change maze learning at all</u>. Now, you do not know this and can only test a sample of the rat population.

Possible Outcomes No. 1

1. You analyze your data with the "proper" statistic and you conclude that the drug Smartamine® really has <u>no effect</u> and thus the null hypothesis (H_O) *is supported*. If in fact Smartamine® really is worthless then you made the correct choice and rejected the alternative hypothesis (H_A). You decide not to test Smartamine® further in terms of maze learning. **Good Job!**
2. You analyze your data with the proper statistic and you conclude that Smartamine® is really great and enhances maze learning; thus the null hypothesis (H_O) *is not supported*. However, in reality, because Smartamine® is worthless you **INCORRECTLY** stated that Smartamine® is going to be the next great learning and memory drug. You conclude that the alternative hypothesis (H_A) is supported. This is a **Type I error** and you are in for a big surprise later on!!

Now let us say that in fact Smartamine® really **does enhance** maze learning, and if you could test all the young adult male rats it would be very clear that this compound is worth millions. Again, you really do not know this but can only sample the rat population.

Possible Outcomes No. 2

1. You analyze your data with the proper statistic and you conclude that Smartamine® works great and the null hypothesis (H$_O$) is not supported, and thus you **CORRECTLY** state that the alternative hypothesis (H$_A$) is supported. You decide to spend a lot of grant dollars further testing this novel drug and visions of the Nobel Prize begin to dance in the corridors of your brain. **Good Job!**

2. You analyze your data with the proper statistic and you conclude that Smartamine® is worthless and that the null hypothesis (H$_O$) is supported, and thus you **INCORRECTLY** state that the alternative hypothesis (H$_A$) is not supported, when in fact it should be supported. You throw your data into the trash and head down to the local watering hole to think up another project. You have probably lost your chance of being a multimillionaire because you made a *TYPE II error,* and will begin to cry when you read how the laboratory down the street discovered the next great learning and memory drug—Smartamine®.

To quickly review: A Type I error, sometimes called the alpha (α) error, is when you say something works and it really does not. A Type II error, sometimes called the beta (β) error, is saying something does not work when in fact it really does.

How is it possible to make these types of errors when you run the proper statistic? And what type of error is worse—a Type I or a Type II? All of statistics is built on probability and holds as one of its basic foundations that nothing is 100% for sure. If the rat population used in the Smartamine® experiments above has a normal distribution, then there will be members of that population that will have a very different reaction to the drug than some of the other rats in the population. For some, the drug will enhance their maze learning ability, and for others it will either have no effect or may even make them worse on any particular day. When you receive your small sample of rats from the breeder, you do not know from which part of the distribution these animals came. While you hope that they are randomly selected and actually correctly represent the normal distribution, they could disproportionally represent one end of the distribution. The sample may contain a greater number of subjects that have a positive reaction to Smartamine®, or by chance you could have gotten a sample of rats that have a negative reaction to the drug. So it all comes down to how much "**faith**" you place in the "**representativeness**" of your sample.

Scientists really do not want to make statements about a great finding only to be proven wrong later on. You might say that it is "bad for business," but in fact it really decreases your credibility. Thus, the scientist states a "*p* value" that will represent significance prior to running an experiment. It goes something like this: If I run my experiment properly and analyze the results with the proper statistics and the results are significant at the $p < 0.05$ level, then I will reject the H_O and state that the H_A is supported. What that actually means is that your results say there are only 5 out of 100 chances that H_O should really be supported. If in fact your results show you actually have a p value of 0.003, it would mean that the chance is only 3 out of 1000 that your <u>sample was not representative of the population</u> and the H_O should be supported. The smaller the p value the greater the chance that you did not commit a **Type I** error. Typically most experiments set their α level at 0.05. If you set your α value too low (e.g., 0.0001) then you may never find significance and thus conclude that there is no effect when in fact there really is, a **Type II** error. So there must be some type of a happy middle ground for all of this. This is where **POWER** comes into play.

Power

Power is **CORRECTLY** rejecting the null hypothesis (H_O) when the H_O is indeed false. Some statisticians like to talk about β, which is a Type II error. Thus power is sometimes defined as 1-beta. If you are willing to take a 20% chance of making a Type II error, then the power is 80%. It is always important to think about the consequences of making a Type II error. Here is an example: Let us say that a DNA sample is collected at a murder scene and later used in a trial. It is the key piece of evidence and DNA is taken from a suspect. If there is no significant difference between the suspect's DNA and that taken from the crime scene, the individual will be convicted of murder and sent to prison for 100 years. One would want to make sure a type II error is not committed (i.e., stating that the two samples were not different when in fact they were).

In the above example with Smartamine®, it was determined that making a Type II error, stating that Smartamine® has no effect when in fact it is a great novel compound, would result in the loss of a financially and academically rewarding outcome. The easiest way to protect against a Type II error is to increase your sample size. But beware, because the larger the sample size the greater the chance of a Type I error (supporting the claim that there is a significant effect when there really is not). You want to have a large enough sample size that adequately represents the population distribution but a small enough α level to have assurance that you support the correct alternative hypothesis.

Statisticians try to advise the experimenter to give him/her the best chance of detecting an important difference if one actually exists. To do this, the statistician looks at not only how many subjects are necessary, but also what is the "magnitude" of the effect the investigator is trying to uncover. To do this, the statistician needs to have an idea of how much variability might be in the data. In a rat maze learning experiment, it may typically take a rat 145 s to complete the maze. If an animal is given Smartamine®, it might be able to complete the maze in 138 s (5% reduction in time). Is that a major effect? Well, you might want to know what the **variance** is (Chapter 3), because that would make a big difference. If in fact the range in times for the Smartamine® animals is only 136–140 s and for the placebo rats it is 142–148 s, then that is a big effect. But if the variance is great, then a mean difference of 138 vs 145 s may be a small effect, and it will require more subjects to show that it is statistically significant. But then it may not be biologically significant.

Statistically significant indicates that differences observed are unlikely to have occurred by chance at some predetermined level. One could say that there is a 1 in 10 chance or 1 in 20 chance that an observed difference was due to "luck." This does not mean that the observed statistically significant difference has necessarily any relationship to any practical difference, although it certainly could. When something has biological significance, it means that the observed measured outcome (e.g., change in weight, change in mobility) makes a truly important fundamental difference in the way the biology of an organism responds to a situation. Biological significance is really unrelated to statistical significance because biological significance is measured in terms of a biological outcome while statistical significance is measured in terms of probability.

Power of a statistic is usually represented as a percent such as 85% power or 50% power. An ideal experiment would have an α level set at 0.01 or lower with a power of 90% or higher. This, however, is extremely difficult to do unless the magnitude of the effect is very large. For most biological experiments, α is usually set at 0.05 and the power at 80%. It is not uncommon to see power at 85% or 90%. One can think of statistical power in the following way: If I were to run my experiment 100 times, what are the chances that I would miss a significant effect when in fact it actually exists?

While increasing the sample size will increase the power of a statistic it also increases the probability of detecting a very small but not biologically significant difference. Depending on the circumstances, sometimes increasing the sample size can be very costly. It is not uncommon to find some studies in which the subject number has been increased just so the authors can claim at least one significant finding.

So what information is actually needed to compute the power necessary for a proposed experiment and how do I get that information? There are a variety of different computer programs available to calculate power and conversely to calculate the sample size necessary for a given power. In order for these programs to work the following information will be necessary:

Calculating Power

1. Population mean of the variable you are interested in
   a. This is usually the value of the control group in your experiment. It is what you expect the normal population to show. For example, in the Smartamine® experiment, it would be the mean time taken for the placebo-treated animals to finish the maze.
2. Sample mean of the variable you are interested in
   a. This is the mean value of the experimental group in your experiment. In our example above, it is the mean time taken for the Smartamine®-treated animals to finish the maze.
3. Variance (**standard deviation**) you observed in the groups
   a. This is the average standard deviation of the two groups.
4. The alpha (α) level that you want to use as the cutoff for significance
   a. Typically this is 0.05 but sometimes 0.01 is required, depending on the importance of the outcome.
5. Number of subjects used in each group
   a. This can get a little tricky depending on the program used. Most of the programs assume an equal number of subjects per group, but others allow an unequal number of subjects. Use the average for the sample size.

Calculating Sample Size

1. Population mean of the variable you are interested in
   a. This is usually the value of the control group in your experiment. It is what you expect the normal population to show. For example, in the Smartamine® experiment, it would be the mean time taken for the placebo-treated animals to finish the maze.
2. Sample mean of the variable you are interested in
   a. This is the mean value of the experimental group in your experiment. In our example above it is the mean time taken for the Smartamine®-treated animals to finish the maze.
3. Variance (standard deviation) you observed in the groups
   a. This is the average standard deviation of the two groups.
4. The alpha (α) level that you want to use as the cutoff for significance

    a. Typically this is 0.05 but sometimes 0.01 is required, depending on the importance of the outcome.
5. The power you want for the experiment
    a. Typically this is 0.80 or 80%.

Key Points to Remember

1. Power is related to the chance of making a Type II error—stating that the H_O is supported when in fact it should be rejected.
2. Increasing the α and protecting against a Type I error decreases power.
3. Increasing the sample size in the experiment increases the power but also increases the chances of making a Type I error.
4. If you are only interested in big effects, then fewer subjects are needed.

One More Word about Power!

You might hear someone say that a parametric statistic (e.g., t-test, Chapter 6) has more "power" than a nonparametric test (e.g., Mann−Whitney U test, Chapter 8) even though they both test the difference between two independent groups. This is indeed the case provided that the assumptions underlying the use of a parametric statistic are valid. One of those assumptions is that the data are normally distributed and another is **homogeneity of variance** (Chapter 6). Homogeneity of variance means that the amount of variability in each of the two groups is roughly equal. If the number of subjects in each group is small then homogeneity of variance is a big issue, but if the number of subjects per group is large (e.g., 20−30) then it tends not to be an issue. The primary reason that parametric statistics have more power is because they use all of the information that is intrinsic to the data. Here is an example:

You are counting the number of astrocytes in a small region of the red nucleus as a function of whether or not the animals are given a drug. Both groups have the same number of animals and were counted independently by the same investigator (Table 2.1). If you analyze these numbers with nonparametric statistics, such as the Mann−Whitney U test, it will show that the two groups are statistically significant at $p < 0.05$ but one does not know by how much. If these same data are analyzed using a parametric statistic, such as an unpaired t-test, not only do we know that the groups are significantly different at $p < 0.05$ but also that the number of astrocytes in the drug group is twice as much as that in the placebo group. We also know that the variance in the drug group is greater than that in the placebo group. For a very enlightening explanation about power see Motulsky.[2]

TABLE 2.1 Astrocytes in the Red Nucleus

	Parametric		Nonparametric		
	Placebo	Drug	Rank	Placebo	Drug
Rat 1	847		1		2687
Rat 3	1107		2		2651
Rat 6	397		3		2397
Rat 9	368		4	1107	
Rat 10	536		5	1005	
Rat 11	827		6		996
Rat 13	496		7		876
Rat 16	1005		8		874
Rat 2		2687	9	847	
Rat 4		1506	10	827	
Rat 5		876	11	536	
Rat 7		2651	12		506
Rat 8		395	13	496	
Rat 12		996	14	397	
Rat 14		874	15		395
Rat 15		2397	16	368	
Mean	698	1548	Median	682	936
SD	285	908			

ONE-TAILED VERSUS TWO-TAILED TESTS

Almost every statistic book has an appendix section that lists several (or in some cases many) different statistical tables, among which is the Student's t-distribution. Depending on the author of that book, that t-distribution table will list values for both a one-tailed and a two-tailed test or maybe just for a two-tailed test. What does this actually mean? Even with a quick perusal of the values in a t-distribution one sees that the critical values necessary for significance are much lower for a one-tailed versus a two-tailed test.

Whenever you set up an experimental design, and you want it to test a specific hypothesis, you have the option to precisely state the direction of the effect in which you are interested. For example, in looking at the effects of a novel compound, you might predict that it is going to change the dependent variable (neuron firing rate) in a positive way (increase the firing rate). But what happens if it has the opposite effect and decreases the firing rate? Would that be an important finding? Maybe it is, but you just do not know for sure, but think it should increase the firing rate. Most statistical tables are set up with the idea that the results could go either way (increase firing rate or decrease firing rate). This is a **two-tailed test** because it looks at both the left and the right sides of the normal distribution. This two-tailed test is sometimes called a **nondirectional test**. The opposite of this is a one-tailed or **directional test** (Figures 2.1 and 2.2).

When do you use the one-tailed values and not the two-tailed values? The criterion for using a one-tailed test versus a two-tailed test is actually dependent on how the original alternative hypothesis is stated and not on the basis of the null hypothesis. First, the researcher must have a solid "hunch" of the direction of the experimental results. This can be obtained from researching the current literature, or from some previous pilot data, etc. Second, the original hypothesis must be specific in stating the direction of the outcome including words such as "increased, decreased, more than, etc." and it cannot be ambiguous. Third, and very importantly, the research hypothesis and directionality must be **stated before any data collected**. Fourth, the researcher must totally ignore any results that go in the opposite direction whether they would be significant or not. Consider the previous research hypothesis:

Research hypothesis: A diet containing 2% of the novel compound Smartamine®, when fed to young adult male Wistar rats (250–275 g) over a 30 day period, *reduces* the number of errors to criterion in the radial-arm maze compared to rats fed a standard lab diet.

This is an example of where a directional test could be run if the investigator chose, **but** they would have to make it very clear prior to collecting data. How do you do this? You simply write the specific hypothesis down in your laboratory record book with a statement that you will be analyzing with a one-tailed test. If the researcher collected data and it showed that the novel compound increased the number of errors to criterion, he would have to ignore those findings. **You cannot go back** and say, "Well I want to now analyze the data as a nondirectional

$$H_o: \mu = \mu_o \quad H_A: \mu \neq \mu_o \quad (\mu \text{ is the population mean})$$

FIGURE 2.1   This is how one writes a statistical statement for a two-tailed test.

$$H_o: \mu \geq \mu_o \quad H_A: \mu < \mu_o$$

FIGURE 2.2   This is how one writes a statistical statement for a one-tailed test.

test." In some situations, it is extremely obvious that the test must be directional. For example, if we are evaluating whether or not a particular type of head trauma will decrease cognitive performance, it is silly to believe that injuring the system will increase cognitive performance. On the flip side, if one is investigating locomotor activity, following a particular type of brain injury, it is quite possible that it could either decrease or increase motor activity. Very often a researcher does not have enough information to be able to actually predict which direction the outcome would be. In these situations only a two-tailed test can be used.

Everyone in neuroscience understands that having statistical significance is much easier to publish than having results that are not statistically significant. Often journals view nonstatistically significant results as negative data. This notion of nonsignificant results as negative data is absolute rubbish and very misguided. Because of this, there is a great temptation to reanalyze data as a one-tailed test when in fact it was originally designed as two-tailed. Who would actually know? Researchers need to be quite cautious about the use of a one-tailed test since it greatly increases the probability of a Type I error. This is because a one-tailed test places the α level in only one side of the sample distribution. Because of the large amount of variance that can occur especially with animal behavior studies, two-tailed tests are usually used.

In a recent experiment, a neuroscientist evaluated whether or not synapse loss occurred in the frontal cortex as a function of aging. Ten individuals were evaluated from eight different decades of life ranging from 20 year olds to 90 year olds. The groups were well matched in terms of time postmortem and gender. As far as the researcher could tell, each of the individuals studied was cognitively normal at the time of death. A drug screen at autopsy showed no significant presence of controlled substances. None of the subjects had suffered any significant neurotrauma including head injury or stroke. The results showed that there was no statistically significant loss of synapses in the frontal cortex as a consequence of aging. **Is this positive data or negative data?** I would definitely say positive data because it means that we probably do not lose synapses in the valuable frontal cortex if we age normally!! Many journals say because there was not a significant age-related difference it is negative data.

BIAS

A researcher was evaluating the effects of a novel compound as a therapeutic agent in his model of traumatic brain injury. As part of the evaluation, the animals were tested in a Morris water maze. If the compound showed a robust effect it was likely that he could obtain additional funding from the army to further develop the therapy. To be unbiased, he had a graduate student from another laboratory inject some of the animals with drug and some with vehicle following the surgical intervention. In this way, both he and his technician would be able to blindly evaluate the animals.

The experimental animals tested in the Morris water maze got four trials per day for 5 days, with each trial lasting up to 120 s (after which they "timed out"). On each trial, the subject started from a different compass location. The task was to navigate the circular pool and find the submerged platform using external maze cues. The dependent measure was latency (seconds) to reach the escape platform. His technician was responsible for entering the data into the spreadsheet and performing the initial statistical analysis. For each animal, each individual trial was recorded using automated visual tracking software. After all the behavioral testing was completed and the animals were killed for histological analysis to confirm injury group, the code was broken and the data were analyzed and plotted. To the investigators dismay, on day 5 (see graph with arrow) the injured drug group was not significantly different from the injured vehicle cohort (Figure 2.3).

But this group of animals was doing so well and was significantly better than the injured vehicle group on day 4 and even better than the

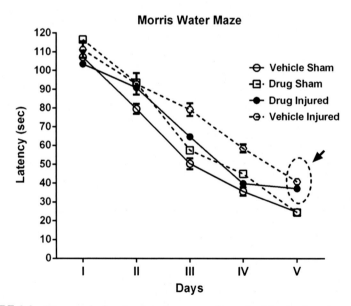

FIGURE 2.3   Line graph showing learning curves for each of the treatment groups. On the last day of training, the drug-injured group appears to have plateaued and is not different from the vehicle injured group indicated by the broken circle and arrow.

drug sham group on that day. What happened on day 5 to change things? He immediately began to investigate the individual trial latency scores for day 5. He found that for several of the animals in the injured drug group, on at least one of the trials for day 5, the animals timed out and received 120 s latency. Typically, in the Morris water maze task, many animals time out during the early learning phase of the task and then as the animals "learn" the task, the latency scores decrease. For a few of these animals, on at least one of the day 5 trials, the animals performed as if they were just learning the maze task (Figure 2.4).

These are important experiments and the future of the laboratory funding may be at stake. One form of bias would be to ignore that animal's trial in which it timed out and only average three of that day's trials. "You know, the animal really 'knows' where the escape platform is located. Besides, something must have happened to that animal on the way to the testing area or in the holding cage immediately before." While it is unfortunate that the results turned out this way, one cannot simply remove scores that do not coincide with a preconceived idea (see Chapter 9).

Bias actually comes in many different forms. Sometimes a piece of equipment malfunctions or is not calibrated correctly. In those situations all the data collected at that time will be systematically off in the same direction in a precise manner. In these types of situations the bias does not contribute to scatter in the data. Investigator bias, which is different, such as cultural in the form of folk truths (toads cause warts) and "conventional" wisdom (he who hesitates is lost) can alter one's view of a data set. There are also areas such as "common sense" and "self-evident truths" where you know something is right because your "intuition" or "gut

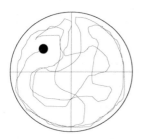 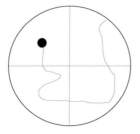 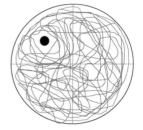

Typical swim pattern on the last trial of day 3. The dot represents the hidden escape platform.

Typical swim pattern on one the last trials of day 4. The dot represents the hidden escape platform.

Unusual swim pattern on one of the last trials of day 5. Animal does not appear to have learned the maze. The dot represents the hidden escape platform

FIGURE 2.4   Computerized tracking of water maze pattern during learning phase.

feeling" tells you so. This is called "truthiness," which was popularized by Stephen Colbert in 2005.

One of the key elements in testing a hypothesis is that the sample one uses must be drawn randomly from the population of interest. That simply means that the collected data are representative of the entire population and not in any way biased. A truly **random sample** really means that if you were to take a small sample, no part of the population would be favored or ignored in the selection process. If you want to know what the most sensitive antibody to synaptophysin is and you only talk to a couple of sales representatives at the neuroscience meeting, it would be a very biased sample, since each company would only highlight their product. The methods used in selecting the sample are extremely critical and differentiate potentially biologically important results from those that belong in the paper shredder. Biased samples are responsible for a vast majority of misrepresented statistical inferences in the scientific literature.

In a nonrandom sample, only individuals with a particular characteristic are chosen from a large group. This can represent a biased sample. How is this different from taking a sample of rats in the vivarium with head injury? In the previous example, the population was defined as all rats with a cortical contusion and no additional distinction was made. In a recent study a new transgenic mouse model of Alzheimer's disease was developed that enhanced the phosphorylation of a specific protein in the hippocampus. Both transgenic and wild-type mice were subsequently screened for their ability to learn a radial-arm maze. Only those transgenic animals that had difficulty learning the task were then evaluated for possible changes in levels of phosphorylation and compared to wild-type mice. The researcher found significantly higher levels of the phosphorylated protein in the transgenic mice compared to the wild-type mice and subsequently concluded that it was the change in protein phosphorylation that resulted in the impaired learning. Since the transgenic animals that did learn the maze were not evaluated, the study resulted in a nonrandom and very biased sampling. This failure to evaluate transgenic mice that did learn the maze would severely taint the conclusions drawn from the analysis.

SUMMARY

- Key to experimental design is stating the null hypothesis.
- The foundation of all statistical tools is probability.
- There are two types of statistical errors and both are equally important.

- A statistical test is only as meaningful as its power.
- Biological significance is not necessarily related to statistical significance.
- One must avoid experimental bias at all costs.

References

1. Cowles M, Davis C. On the origin of the .05 level of statistical significance. *Amer Psych*. 1982;37:553–558.
2. Motulsky H. *Intuitive Biostatistics*. New York: Oxford University Press; 2014.

Statistic Essentials

Stephen W. Scheff

University of Kentucky Sanders—Brown Center on Aging,
Lexington, KY, USA

It's a capital mistake to theorize before one has data. Insensibly one begins to twist facts to suit theories, instead of theories to suit facts. **Sherlock Holmes**

Fundamental Statistical Principles for the Neurobiologist
http://dx.doi.org/10.1016/B978-0-12-804753-8.00003-8

37

TYPES OF DATA

There are four different types of data (NOIR) and each is associated with a specific set of rules on how they should and should not be presented and analyzed. The four different types are Nominal (N), Ordinal (O), Interval (I), and Ratio (R). Let us first describe each of these different types of data.

NOMINAL DATA

Nominal data, also known as **categorical data** or **classificatory data**, is descriptive or qualitative in nature. One simply counts the number of times a particular event takes place. A typical example would be the type of rat or mouse used in the laboratory. We could have different types of rats: Sprague—Dawley, Fisher 344, Wistar, Hooded, etc. In fact one could count the number of these different rats in the university/research vivarium. We could even assign each different type of rat a number (e.g., Sprague—Dawley is a 1; Wistar is a 2). What is unique about this type of data is that even though we assign numbers to the different species of rats, they are only descriptors and do not have any actual numerical meaningfulness. The numbers do not have any quantitative value. We could just as easily assign them letters such as A, B, C, etc. By assigning a descriptor one can only say that the individual members of the data set differ from each other with respect to some specific characteristic (e.g., Hooded rats are black and white while Sprague—Dawley rats are all white). There is no relationship between these characteristics. The assignment of a number to the different species is totally random and it does not mean that Sprague—Dawley rats are lower or higher or better or worse than Wistar rats. In fact, what a nominal scale really represents is equivalence. If a number of given rats is assigned to the Wistar group then each of those rats is considered to be similar to other Wistar rats. The only restriction is that the nominal variable consists of a set of alternative and

TABLE 3.1   Rat Species in Different Rooms in the Vivarium

	Vivarium location				
Species	Room 5867	Room 5872	Room 5916	Room 5924	Room 5963
Sprague–Dawley	8	15	24	6	17
Fisher 344	7	21	9	10	26
Wistar	5	9	32	3	5
Total	20	45	65	19	48

mutually exclusive categories, and any given observation, in this case a rat, cannot be in more than one of the groups.

Assigning a numeric identifier to a particular group (e.g., Wistar = 2) does not mean that one can carry out some mathematical function such as adding, subtracting, dividing, or multiplying these nominal numbers. It is important to understand that the number associated is simply an identifier and could be replaced by a star or triangle. If the experimenter assigns a number as an identifier to nominal data, and if the numbers are used in a logical sequence, then one knows how many groups are used in the experiment.

Can nominal data be statistically analyzed? Definitely if the data are in the form of a proportion or percentage and the statistic is a chi-square test. For example, in a group of rats that we found in room 5867 of the vivarium, there were 8 Sprague–Dawley, 7 Fisher 344, and 5 Wistar. Is this proportion of rat species typical of the normal population in rooms that have rats? One could compare the distribution in room 5867 with any of the other rooms or with the total of all rooms (Table 3.1). As another example, one could also count the number of different neurons in a given region of the cortex (e.g., stellate, large pyramidal, small pyramidal, Martinotti, chandelier) and compare different cortical regions to see if they have the same proportion of neurons.

Nominal or categorical data can be analyzed with a chi-square test (Chapter 8) or a binomial test. There are also several different tests, such as the Cramer coefficient, that can be used to assess the possible association between categories.

ORDINAL DATA

Ordinal data are also qualitative, but unlike nominal data, the variables have a specific order or rank that is related to the alternative amounts or degree of intensity of the variable. In this regard they are a little more complex than nominal data. The numbers associated with an ordinal scale need to be applied systematically so that a rank of 3 follows a rank of 2, which follows a rank of 1. However, the quantitative distance between any two of the ranks has no particular value. A good example would be rating male mice on aggressive behavior. One could assign overall levels of aggression on a 1—5 scale with 1 being the least aggressive and 5 being the most aggressive. The only scores a mouse could receive are a 1, 2, 3, 4, or 5. There is no such thing as a 1.5 or a 2.3 on this scale only whole numbers (1—5). Another way of thinking about this is, you can assign a letter of the alphabet to each category and it still makes sense. (e.g., 1 = A, 2 = B, ... 5 = E). Mice with 4's and 5's are more alike than 3's and 5's. But one cannot say how much more a 4 is than a 3, or what is the difference in aggression between a 1 and a 3, only that a 3 is more aggressive than a 1 or a 2. From an experimental design aspect, the experimenter must have a strict criterion for each of the different categories prior to the start of the observation.

Often we are asked to rank something (different craft beers) on the scale of 1—10 and then you are never told if 1 is the best or worst craft beer you have ever tasted. So whenever an ordinal scale is used it is imperative that the value of the rank be made obvious. In the example above, A or 1 = least aggressive and E or 5 = most aggressive.

A very common use of an ordinal scale is in the field of Alzheimer's disease (AD). One index of pathology is the number and distribution of neurofibrillary tangles in the neocortex. An anatomist, Heiko Braak,[1] came up with a classification scheme, using the ordinal scale of 0—VI, to describe the degree of AD pathology. He purposely used Roman numerals to indicate the degree of pathology. It is now routine for a neuropathologist to report that a particular patient at autopsy had a Braak score of III or IV. Having a Braak score of IV does not mean twice as much pathology as a Braak score of II but rather that the distribution of the pathology is simply more expansive. There are many scientific papers in the AD literature that will report something like, "The mean Braak score was 5.2." This is totally impossible since this ordinal data (like the nominal data) cannot be subjected to arithmetic operations. Many investigators will tell me that they can put Braak scores into their statistical programs and obtain a mean value with a standard deviation. I am simply very curious how these programs are able to read Roman numerals.

Can ordinal data be statistically analyzed? Absolutely and some typical examples are the **sign test, Mann—Whitney U test, and Kruskal—Wallis test** (see Chapter 8). Most statistical software packages have a limited number of routines that can be used to analyze ordinal data. It is probably quite obvious that both nominal and ordinal data result in very limited information and most neuroscience types of research should try to avoid these data types. If it is important to report a measure of central tendency using an ordinal scale, then reporting the mode would be the best choice, because it only involves counting. *Reporting the mean is totally inappropriate when using nominal or ordinal data.*

INTERVAL DATA

Interval data are quantitative and use a fixed unit of measurement. This type of data differs from both nominal and ordinal in that one can apply mathematical operations to it. One can think about this data type as combining both nominal and ordinal data with information about equality. The distance between sequential values is the same. For example, in a learning experiment rat No. 1 makes 6 correct choices and another rat (No. 2) makes 12 correct choices. We can say that rat No. 2 made twice as many correct choices compared to rat No. 1. What is absolutely critical in using interval data is that the intervals along the scale are equal to each other. Interval data do not have a set zero point since what is more important is the numerical distance between values. A very common example is changes in temperature. A solution that changes temperature from 37 °C to 39 °C is the same amount as another solution that changes from 48 °C to 50 °C. **Can interval data be statistically analyzed?** Absolutely and a good example is the **Fisher test** or the **Fisher–Pitman test**.

RATIO DATA

Ratio data are the highest order of quantitative data and are characterized by the presence of an absolute zero. So when something is said to be zero it means nothing is there. In true ratio data it is impossible to have negative numbers. It should be obvious then that measuring temperature in Centigrade and Fahrenheit is interval data. If you want to measure temperature using a ratio scale then it would have to be on the Kelvin scale, where the reading of 0 signifies the lowest possible temperature, because it is the temperature were all molecular movement stops. Of course we thank William Thomson (aka Lord Kelvin), a British inventor and scientist, for that. Most scientific data are of the ratio type and they can have infinite levels, although in most cases there is usually an upper limit. For example, in a given nucleus within the thalamus there are normally no degenerating synapses. However, following a lesion in the brainstem there may be many degenerating axons that project to the thalamus, resulting in a large number of degenerating synapses. The number of course would be the result of the severity of the injury and the time post trauma. A major difference between ratio data and interval data is that ratio data usually have a normal distribution while interval data usually do not. There are many different statistics that can be used to analyze ratio data (e.g., **t-test, analysis of variance**). Most statistics books and commercially available software programs have many different tests to analyze ratio data.

All data can be assigned to one of the four scale classifications and these four scales have the following levels of precision:

Nominal < Ordinal < Interval < Ratio

The most precise, ratio data, includes all the characteristics of the other three. Interval and ratio data (sometimes referred to as **cardinal numbers**) are much more informative than nominal and ordinal data.

DISCRETE AND CONTINUOUS DATA

What is meant by the terms **discrete** and **continuous data**? Both interval and ratio data can be considered as either discrete or continuous. A discrete number is one that is clearly different from the one next to it. Let us say you are counting the number of rats in two different rooms in the vivarium. Since you cannot have a half a rat, the number you count is discrete. However, if you are getting the average weight of rats in the two rooms, one average might be 276.4 g and the other 212.8 g. A good way to think about discrete versus continuous is whether or not the units of measure can be subdivided multiple times and it still makes sense. For example, a meter can be subdivided infinitely (centimeter, millimeter, micrometer, nanometer, etc.) and thus is considered continuous. Fortunately, almost all statistical techniques work with both discrete and continuous data.

Remember to Avoid This Statistical Faux Pas

One of the major mistakes that is commonly seen in the neuroscience literature is assigning cardinal scores (1, 2, 3, ... N) to ordinal data and then applying arithmetical operations to these numbers. Using the example above when studying aggressiveness of a rat following a drug injection, there are five different levels: very aggressive (5); moderately aggressive (4); mildly aggressive (3); not aggressive (2); afraid (1). The experimenter then makes the statement *"Rats on drug A had a mean aggression score of 3.8, while rats on drug B had a mean aggression score of 2.9. Thus, drug A made the rats 30% more aggressive."* This type of statement may be true but one cannot state it using ordinal data.

MEASURES OF CENTRAL TENDENCY

Arithmetic Mean

The most common measures of central tendency are the mean, median, and mode. Everyone in science already understands that the **arithmetic mean** is an average value of a group of data points. This is what is normally called the sample mean $(\overline{X})$ as opposed to the population mean (μ), which is the theoretical value that would be obtained if every member of a population were assessed. When the arithmetic mean is calculated it does not necessarily follow that any of the actual data points are identical to the calculated mean or even necessarily close to it. This is because values that are outliers can skew this value. The arithmetic mean alone does not really provide much information by itself and needs to be accompanied by some measure of variance such as the standard deviation. There are two other means that are sometimes used in statistics: **geometric mean** and **harmonic mean**.

Geometric Mean

In mathematics, the geometric mean or average also signifies a measure of central tendency but it is calculated very differently. It is the average of the logarithmic value of a data set that is converted back to a base10 number. The value of doing this is that it tends to negate the significance of very high or very low numbers. One simply finds the log value of each of the data points and then determines the average log value. The geometric mean is the antilog value of this average. It is used sometimes in statistics to make data more meaningful. For the most part this is not a measure of central tendency that is used routinely. The bottom line is that it differs dramatically from the arithmetic mean and is not a simple average. For example, take the following density of some specialty cells in a given region of the amygdala:

$$14/mm^2; \ 97/mm^2; \ 27/mm^2, \ 437/mm^2, \ 134/mm^2$$

The arithmetic mean is 142 and the geometric mean is 74.5.

Harmonic Mean

In statistics, the harmonic mean is used when there are unequal sample sizes especially when applying post hoc tests after an analysis of variance. A number of different post hoc tests (Chapter 6) require that the number of subjects in each group be the same. For example, both the honestly significant different test and the Newman–Keuls test require an equal number of subjects. If in fact the n's are not equal, and do not differ substantially, then many statistical software programs will calculate the harmonic mean. The reason for this is not very complicated. If you have very few subjects in one group and many in another group, then the variance in the smaller group has a much greater influence than the larger group. What a harmonic mean does is provide a special average sample size. This special average is obtained by dividing the number of groups by the sum of the reciprocals of the different group sizes. Depending on the number of groups the calculation can become complicated, but fortunately the computer software does all the work. Many of these programs will indicate that the harmonic mean was used. The harmonic mean technique is sometimes also referred to as an **unweighted means** technique.

Weighted Mean

A weighted means technique is used in statistics when the sample size between groups is unequal. Instead of each data point contributing equally to the statistic, individual group means are adjusted depending on that particular sample size compared to the overall sample size. For example, if the total sample size is 36 and the different group sizes are 7, 10, 9, and 10, the groups with 10 should be given more influence than the group with only 7. In other words, the different group means only contribute in direct proportion to their sample size. Some commercial software programs use this approach when dealing with unequal sample sizes, which tends to make the F values in a one-way ANOVA smaller. Overall, it does not really affect the outcome very much but statisticians talk about it. It is used most often in descriptive statistics.

Median

This is simply defined as the middle-most score in a group of numbers. One simply rank orders the data points within a group from lowest to highest or highest to lowest and then determines which value lies in the middle. It is very rare that the mean and the median are identical, and the median is usually lower than the mean but not always.

If there is an odd number of values then it is extremely easy to obtain the median. Simply rank order the data values and find the middle value:

28, 39, 45, 48, 62, **69**, 77, 85, 93, 94, 97   median = **69**

It becomes a little more difficult when there is an even number of data points because the median falls between two scores:

39, 45, 48, 62, **69, 77**, 85, 93, 94, 97   median = **73**

Suppose the values were 39, 45, 48, 62, **70, 77**, 85, 93, 94, 97.

One would still take the two middle values (**70** and **77**) and add them and divide by 2. The median would be **73.5**. The median is more informative as a "stand alone" value compared to the arithmetic mean because you know that half of values of the data set are below and half above this value. One of the major drawbacks to using the median is if the data set consists of a very large set of values (e.g., 75 or more). You would have to rank order this large set of values. Another disadvantage is the fact that if the middle-most number moves even slightly then the median would be altered. The arithmetic mean, however, is relatively unaffected by a change in almost any of the middle numbers of a data set. However, the mean is easily affected by an extreme value at either end of the rank ordering while the median is not. The median value is extremely important in nonparametric statistical procedures (Chapter 8) and is used extensively in describing biological data because it is relatively immune to biological variance. Biological variance refers to differences that can occur simply because the subjects developed differently. At any given time, no two mice will have exactly the same number of astrocytes in the substantia nigra even if they are litter mates. The difference in astrocyte number in two identical mice is considered to reflect biological variance.

Mode

This descriptive statistic is used extensively to describe frequency distributions. It means the most frequent (some would say most fashionable "à la mode") value in a data set. For example, in the data set: 44, 44, 62, 91, 121, 121, 121, 164, 173, 189 the mode is **121**. This is because the value 121 occurs more often than any other value in this set of values. If in fact each of the values is unique and appears only once, then the data set does not have a mode. If a data set has two modes, then it is called **bimodal**. In Figure 3.1 there is a bimodal distribution of neurons with one mode in the 13–17 μm range and the second in the 43–47 μm range.

One of the main strengths of the mode is its ability to indicate whether or not a set of data points has a normal or unusual distribution. Data sets that do not conform to a normal distribution cannot be used with many parametric statistics. It can be very informative in determining why the results of a particular experiment do not "appear" to have significance. The mode is a very useful statistic if the distribution of the data set is a skewed distribution. What this means is that it is a lop-sided pattern when graphed. For example, if the values of the distribution of neuron size had a normal distribution (Figure 3.2) it would have a mean, median, and mode that were essentially the same.

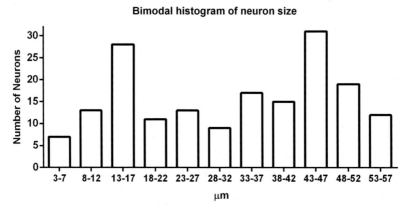

FIGURE 3.1   The overall range of neuron size is 3–57 μm in diameter with most of the neurons clustered around 13–17 and 43–47 μm. This grouping was arbitrarily defined by the experimenter.

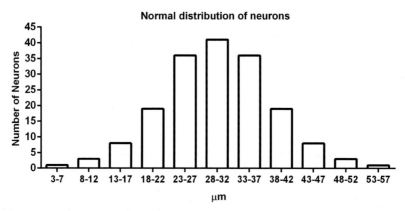

FIGURE 3.2   This is a totally normal distribution of neurons that would be extremely rare unless a very large number of neurons were sampled. It is normal because it is exactly symmetrical with the mean, median, and mode exactly the same.

Statisticians will often talk about data sets that have a skewed distribution. This does not mean that the data cannot be analyzed but it does indicate that most of the variance falls within either the top or bottom half of the distribution. As one might expect, statisticians definitely prefer to work with normal distributions.

A positively skewed distribution has a majority of the data values at the left side (Figure 3.3). The mode for this distribution is also shifted to the left while the mean is shifted to the right.

Just the opposite is the case for a negatively skewed distribution (Figure 3.4).

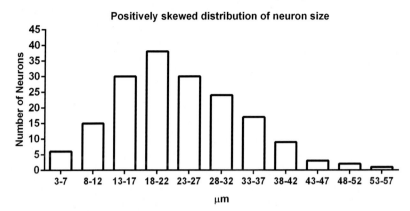

FIGURE 3.3   The entire distribution of cells is shifted to the left side of the graph.

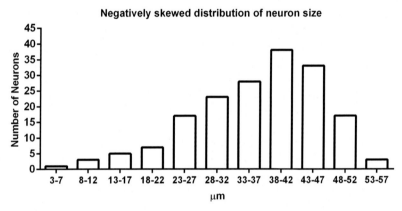

FIGURE 3.4   As opposed to a positively skewed distribution, the negatively skewed distribution has the majority of values shifted toward the right side of the graph.

So what is so important about a normal distribution? For all practical purposes it has a bell-shaped curve and its main features are that the three measures of central tendency (mean, median, mode) all lie at relatively the same place on the curve (have almost the same value). Johann Carl Friedrich Gauss, an eighteenth century mathematician, is the individual responsible for coming up with the normal distribution. Consequently the normal distribution is called the **Gaussian distribution**. Although it is hard to believe, most biological data tend to fall into a normal distribution. A true Gaussian distribution is symmetrical and the tails never actually touch the X-axis. While actually obtaining data that is a perfect bell-shaped distribution is extremely rare, how much can it actually deviate before is it no longer considered to be "normal"? Most statisticians simply look at a data set, or the shape of the data when graphed, and use their "educated" guess to decide. For the most part it is pretty easy, especially if there does not appear to be a bimodal distribution and there are not many very "deviant" values. Alternatively there are some actual statistics, such as the chi-square statistic, that can be used.

VARIANCE

One of the primary goals of using statistics is to evaluate a sample and then make some statement about relevance to the much larger population from which that sample was randomly drawn. For example, if you sampled 10 Sprague—Dawley rats from the vivarium and found that they weighed on average 310 g, how confident are you that the next 10 Sprague—Dawley rats you weigh will also be 310 g? The sample of rats that you weighed is indeed a type of estimate of the entire population of Sprague—Dawley rats in the vivarium but how good of an estimate is it? Well, it is what you would expect if someone performed the sampling the same way you did. It is quite possible that the sample you took was biased and really does not reflect all the Sprague—Dawley rats in the vivarium. Perhaps you only weighed animals from one room and all of those rats were on a special diet or maybe you happen to get all littermates. Here is another example: If you test a drug on the firing rate of neurons in the substantia nigra and find that it increased the firing rate by 28%, how confident are you that someone else in the lab (or in another lab) that tests this drug will also find a 28% increase in firing rate? Regardless of how careful you were in weighing the rats or in evaluating the neuron firing rate, you would find that there were some differences between rats and some differences between neurons in your sample. This difference is called variance. Statistically, variance refers to some type of random error. If the error is not randomly distributed then it comes from biased sampling, regardless of whether or not the techniques used were very precise. Biological variance accounts for the greatest portion of variation in a set of data. While there are certainly other contributors to the variance in a set of data (e.g., equipment imprecision), biological variance is the major source.

Golf is an expression of motor skills and cognitive ability and shows a lot of biological variance. Here are the golf scores for two different players after they played multiple 18 hole rounds on the same course. If a golfer made par on every hole he would score a 72 (Table 3.2).

If you look at these two sets of scores you can see that they both have the same mean score for a round but there is considerable variability in the different data sets for each golfer. In fact, Golfer No. 1 appears to be far more consistent, while Golfer No. 2 might have some good rounds, but the "good" rounds are offset by his numerous not so fantastic ones. One

TABLE 3.2   Golf Scores for Two Different Individuals

Round	Golfer No. 1	Golfer No. 2
1	84	86
2	90	83
3	86	96
4	86	91
5	85	88
6	88	82
7	87	93
8	86	92
9	91	79
10	85	93
11	88	80
12	85	78
Mean	**87**	**87**

could say that there is more variance in the golf play of Golfer No. 2 compared to No. 1, and that Golfer No. 1 may be more precise. If Golfer No. 1 were very precise, she would make an 86, 87, or 88 on every round. Predicting what score she will have on the next round is in part what statistics tries to do.

There are numerous ways to describe variability in a set of data. One of these is called the **mean deviation**. This is simply taking the absolute difference (ignore the negative signs) between a given score and the mean score and dividing it by the total number of scores. Why not just use the sum of the mean differences? If you did this, and retained the positive and minus signs, the result would be zero. So instead, you divide it by the total number of scores. For Golfer No. 2 above, the mean deviation score would be 5.5. This mean deviation is very seldom ever used because it ignores negative values that could be very important. What statisticians commonly do is perform a **Sum of Squares** (SS) measure, which simply means the deviations from the mean are squared and then these squared numbers are added together. Why do they do that? With this technique, the negative values are still retained in the data set. This Sum of Squares measure is actually a representation of the dispersion of the data and it is used in a many different statistical formulas including some very common descriptions of variability, the variance (S^2) and the standard deviation (**SD or sd**) (Table 3.3).

TABLE 3.3 Difference between Golfers

Round	Golfer No. 1	Difference	Absolute difference	Golfer No. 2	Difference	Absolute difference
1	84	−3	3	86	−1	1
2	90	3	3	83	−4	4
3	86	−1	1	96	9	9
4	86	−1	1	93	6	6
5	87	0	0	88	1	1
6	88	1	1	82	−5	5
7	87	0	0	93	6	6
8	86	−1	1	92	5	5
9	91	4	4	79	−8	8
10	85	−2	2	93	6	6
11	89	2	2	81	−6	6
12	85	−2	2	78	−9	9
Mean	87			87		
Total	1044	0	20	1044	0	66
Mean deviation			1.67			5.5

Statisticians talk about the measure of variance a lot, and in fact it is at the very essence of statistics. So what is statistical variance? It is actually defined as the mean of the Sum of Squares or the mean of the sum of the squared deviations from the mean. All one has to do is obtain the Sum of Squares and divide by the total number of values (n) in the data set, which is represented by the symbol S^2. Statisticians use a capital letter to signify that this is a population statistic (Chapter 2). But wait a minute, since one almost never knows the true mean of a population, how can you calculate the population variance? Since we are only evaluating a small sample, any estimate of the population variance will be incorrect because the scores in our sample will deviate less from their own mean than the population mean. This is a great question and in fact statisticians have a way around that (of course). To obtain an unbiased estimate of the population variance, the statisticians divide the Sum of Squares by n−1 (instead of n), which is the sample size minus one. What this does is force the variance to be larger, essentially making it a more conservative estimate. This is then represented by s^2 (note it is a small s). Thus, whenever you see s^2, you will know that this is the variance associated with your data sample. In our example above, the s^2 for Golfer 1 is 4.55 and for Golfer 2 it is 39.82. Again notice the big difference in variance between the two golfers.

STANDARD DEVIATION

The **standard deviation** (**SD**, also abbreviated **sd**) is perhaps the most widely used measure of variability in a data set. To calculate the SD all you do is take the square root of the s^2. If that is the case, then the SD of a sample could also be represented as a small **s**, and in fact it is in many of the statistical formulas. So if we know the mean and the SD, and we assume that the data are normally distributed, then we have information about the entire distribution of scores. So why use the SD and not the s^2? In essence, the SD tells us how much the sample scores actually deviate from the sample mean. Looking at the two golfers it is now quite obvious that the overall scores for Golfer No. 2 are approximately $3\times$ as variable as Golfer No. 1.

Very often in a manuscript one will find that the mean and SD are presented: 87 ± 2.1 (Table 3.4).

TABLE 3.4 Mean and SD Golf Scores

	Golfer No. 1	Golfer No. 2
Mean	87	87
SD	2.1	6.3

Hypothetical Normal Distribution

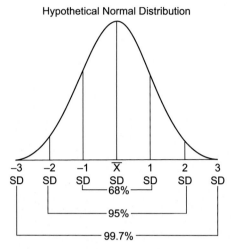

FIGURE 3.5   This normal distribution is actually the foundation on which most of the parametric statistics are based. One standard deviation (SD) to the left and the right of the $\overline{X}$ represents 68% of the area under the curve. At two SDs to the left and right, the area under the curve is 95%. That really means that 95% of the values that make up the normal distribution are contained within −2 SD and +2 SD.

In a normal distribution, 68.2% of the data points fall between −1 SD and +1 SD of the mean and 95.4% between −2 and +2 SD and 99.7% of the cases within 3 SD (Figure 3.5).

The bottom line is that the SD actually describes the distribution of individual scores around the sample mean. The SD can only be used with the mean and does not have any meaningfulness in regard to the median or mode. In fact, reporting the SD with the median is simply incorrect. Just like the mean, the SD is extremely sensitive to outliers. If in fact Golfer No. 1 had a shoulder injury just prior to the 12th round of golf and shot a 120 instead of an 85, her mean score would become a 90 and the standard deviation 9.7. Even though the SD is simply the square root of the sample variance, the reason for reporting the SD is that it is in the same units of measurement as the mean. Every commercially available statistical program, and many hand-held calculators, will automatically calculate the SD for a group of data.

STANDARD ERROR OF THE MEAN

Let us go back to the question of the weight of Sprague–Dawley rats in the vivarium. You sampled 10 Sprague–Dawley rats from the vivarium and found that they weighed on average 310 g. How confident are you that the next 10 Sprague–Dawley rats you weigh will also be 310 g? One way to do this would be to resample the Sprague—Dawley rat population many, many times (like maybe 20) and of course making every Sprague—Dawley rat in the vivarium available for sampling. With a little bit of effort we could then find the mean of all these samples, essentially getting the **mean of the means**. Of course we could then calculate the standard deviation of this mean of means. This new standard deviation is called the **standard error of the mean (SEM)**. So to obtain the SEM you really have to replicate your study multiple times (cannot really tell you how many, but a lot). However, your friendly neighborhood statistician has come up with a calculation of the SEM based on the SD from a single sample. Essentially what is done is to divide the SD by the square root of the "n". What this will do is immediately make the measure of the variance much, much smaller. As the number of subjects increases, the variance will shrink even more. So if you want to make your experiments look really good you report the mean ± SEM. *But is that really correct?* Actually **it is not correct** because the SEM is really a population statistic. It states the variability around a set of means when in fact you only have a small sample of individual data points. It is important to remember that the sample SD is not the equivalent of the population SD and it may not even be close but it is a better predictor than the SEM.

So why does the statistical software package give the SEM when a set of values are entered in? The commercial computer program has no way of "knowing" from where your data came. It could have come from multiple replications of your experiment. It is unfortunate that many statistical programs use the SEM as the default measure of variance in a set of data. It is important to check first before graphing these results. However, the SEM may be very important for other statistical measures.

CONFIDENCE INTERVAL

What a confidence interval (CI) does is give you a range of values that provides some level of assurance that those values will contain the mean value of **the population** you are interested in. Let us go back to the sample of rat weights that you got from your sample taken from the vivarium. The mean of your sample was 310 g, but since not all of the rats weighed that, there had to be some variation. You can compute that variance and report it is as the SD, as discussed above. Thus you would report that the average weight of the Sprague—Dawley rats in the vivarium was 310 ± 80 g. That ±80 is the estimate of some type of margin of error. The combination of the ± SD is your best guess (ok estimate) as to the weights of most of the Sprague—Dawley rats. That SD after the mean signifies that from your sample you are willing to predict that there is a 68% chance that anyone who weighs a Sprague—Dawley rat today using your method of choosing a rat in the vivarium (which may not be as random as you intend) will find it weighs between 230 and 390 g. How can we say that? Because 1 SD is 68.26% of the normal distribution and we are assuming that the weights of the Sprague—Dawley rats in the vivarium are normally distributed.

If you are ever bored while in the laboratory and looking for something to waste time on, copy this normal curve onto a piece of cardboard and cut out the normal curve. Weigh this cut out piece of cardboard. Now cut out the section that is indicated by 1 SD to the left of the mean. If you reweigh the cardboard normal distribution without the 1 SD piece, you will find that it is exactly (provided you can cut accurately) 34.14% lighter. This will be the case regardless of whether or not the normal distribution has a very sharp peak or has a relatively low peak such as the normal curve below (Figure 3.6).

However, you really do not know if your sample was a good estimate of the total population of rats in the vivarium and besides your sample was probably pretty small compared to the entire rat population. What you really need is a better estimate. That better estimate is called the confidence interval.

Normally when the CI is reported it is given by a pair of numbers with a dash in between such as 275—345. The 275 is the lower end of the CI and the 345 is the upper end. These two numbers make up the upper and lower **confidence limits**. It is important to understand that the <u>CI is a representation of the population</u>. As we saw in the previous section, the

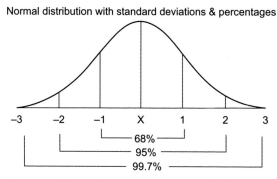

FIGURE 3.6   In this normal distribution the vertical lines represent standard deviations. Note that one standard deviation on either side of the mean (X) when added together represents approximately 68% of the area under the curve. When you increase this to 2 standard deviations on each side it represents approximately 95% of the area under the curve while 3 standard deviations is 99.7%. Note that this is still a normal distribution although it is flatter than the one in Figure 3.5.

SD is great for a sample but the **SEM is a better estimate of the variance associated with a population**. However, you have to remember that the SEM is still related to your sample and simply gives you an estimate of what the variance might be if you were to run your experiment a number of times. Just like the SEM, as you increase the sample size used to compute the CI (and the SEM), the CI range will become smaller.

Let us say that you wanted to be 95% or 99% confident of the range and what does that actually mean? The CI is your best "guestimate" of a range of values for the population (here all of the Sprague–Dawley rats in the vivarium) derived from your sample. Most often we use 95% because then we are saying there should only be about a 5% chance that value will fall outside of this range due to random sampling. One would say that the **confidence level** is 95% or 99%. Whenever you use a confidence interval you have to state the confidence level and thus you would write that the 95% CI = 275–345. The goal is to present a CI that is as narrow as possible that still has some guarantee that the true value of the population falls within the confidence levels you specified.

If you go to a table of normal distribution you can actually look up what are called z scores that represent a given area of the curve of a normal distribution which is then related to the standard deviation. Here is an example: If you were to look up the value of +1 SD, which we already know is equivalent to 34.1% of the area under the curve, the z score (value) would be 1.00 (Table 3.5).

For a 95% probability, the z score would be 2 × **1.96** because that represents 95% of the area under the curve of the normal distribution. So the limits of the CI would be −1.96 and +1.96 SDs. Calculating this is really quite simple and the SEM is used instead of the SD (so it is important to know the SEM also). Here are the weights of the Sprague–Dawley rats in your sample. Note how much lower the SEM is than the SD.

TABLE 3.5 z Score and Area under the Normal Distribution

z score	Area	z score	Area	z score	Area
0.95	0.3289	1.45	0.4265	1.95	0.4744
0.96	0.3315	1.46	0.4279	**1.96**	**0.4750**
0.97	0.3340	1.47	0.4292	1.97	0.4756
0.98	0.3365	1.48	0.4306	1.98	0.4761
0.99	0.3389	1.49	0.4319	1.99	0.4767
1.00	**0.3413**	1.50	0.4332	2.00	0.4772

To get the 95% CI for this sample simply multiply the z value for the 95% probability (1.96) by the SEM (25.3) = 49.6 and both add and subtract it from the sample mean

$$95\% \ \text{CI} = 310 \pm 49.6 = \textbf{260.4} - \textbf{359.6}$$

But there is a catch to all of this. This actually works great if you know that your sample variance is a good reflection of the population variance. In most practical situations you have no idea of the true population variance. The other factors that can influence your CI is the sample size and the population size. In most experiments the sample size is small and you do not know the population standard deviation. In this scenario you actually have to use a different procedure and instead of using the z score to figure out the CI you use a t-distribution. Also, instead of using a sample size of 10, one uses n−1 (in this case 9) as the sample size since it will more closely reflect the population. Fortunately, every commercially available statistical package uses this technique to calculate the CI. In fact, if you go on the web you will find that the CI calculators (and there are several) will automatically calculate the CI for you and they use the t-distribution coupled with n−1.

For our experiment, looking at the data in Table 3.6, the commercial calculators report the 95% CI as 253−367. How do you interpret this CI?

TABLE 3.6   Sprague−Dawley Rat Weights in Vivarium

Rat	Weight (g)
1	187
2	312
3	384
4	159
5	298
6	351
7	373
8	334
9	398
10	304
Mean	310
SD	80
SEM	25.3

TABLE 3.7 z Value and t-Distribution Values for Sample Size of 10

Confidence Interval	Area under curve	z value	t value[a]
80%	0.400	1.28	1.38
90%	0.450	1.64	1.83
95%	0.475	1.96	2.26
99%	0.495	2.58	3.25

[a]Values reflect n−1.

You are 95% confident that if the mean weight of Sprague–Dawley rats in the vivarium on that particular day were known, the interval 253–367 g would contain it. The CI really pertains to the interval containing the mean rat weight and not whether the mean weight is actually in the specified interval (Table 3.7).

The CI can also be calculated for the differences in means. In brief, one first calculates the pooled estimate of the population variance, which is calculated from the SD of the two groups. Then you determine the SEM of the pooled variance, and use it to calculate the CI at a specific level (e.g., 95%). This value is then added and subtracted from the mean difference of the two groups.

There are several very good journals that now require that the authors present both the SD and the 95% CI when reporting results. There are also journals that only want the CI in place of any normal significance testing. If you are submitting to that type of journal you need to consult with your local friendly statistician because they will want to discuss with you the amount of your effect size that you believe should be significant. A discussion of this should be held with your biostatistician and is beyond the scope of this book.

STATISTICAL MYTH CONCERNING CONFIDENCE INTERVALS

Many researchers erroneously believe that if the CIs of two groups overlap, the group means cannot be significantly different. This is certainly the case if there is a very large overlap, but actually they can overlap by as much as 29% and still be significantly different. There can also be considerable overlap with the standard deviation and the group Means Can Still Be Statistically Different.[2]

WHAT IS MEANT BY "EFFECT SIZE"?

Effect size is the magnitude of difference between groups or sometimes called the magnitude of the experimental effect. For example, let us say you are testing a novel compound on cell firing rate in the red nucleus. When an animal is treated with the novel compound they show an average of 15% increase in firing rate over baseline. This really does not say anything about the variability in the change in firing rate within the group receiving the novel compound. But then you have to ask the question "Is a 15% change in firing rate really that meaningful?" Let us say that when you treated with vehicle there was a 5% change in firing rate. Now the difference between the novel compound group and the group treated with the vehicle is 10%. So it looks like the effect size is 10%. You then run a simple statistic to look at the difference between groups and yes there is a significant difference ($p < 0.05$). If you are a graduate student and you find significance you are thrilled because your experiments are working and you might even graduate!

What one really wants to know is whether or not a 10% difference in cell firing rate is really all that meaningful. Let us put this in terms of the null hypothesis. The null hypothesis would say that there is no difference between the groups and thus the effect size is zero. However, with even a little bit of difference between groups the effect size would not be zero. But is that little bit of difference really meaningful. There is a nifty statistic called **Cohen's d** that was developed by Jacob Cohen.[3] This statistic is used to indicate the effect size of the differences between treatment means, which is often called the **standardized mean differences**. Another statistician, Larry Hedges proposed a slight modification of Cohen's d and it is called **Hedges' g**.[4] It is a little more conservative (some would say a little more accurate) than Cohen's d but interpreted exactly the same. These statistics take into account the differences between the means in question, the pooled variance in the two groups, and the size of the groups used to calculate the means. The values of d or g can be equated to the standard deviation. For example, if the d score is 0.8 it means that the two groups differ by 0.8 standard deviations. A score of 1.4 would be 1.4 standard deviations. What Cohen did was set up values for what he considered to be "large," "medium," and "small" effect sizes. A small effect size ($d \leq 0.2$) indicates a difference that is somewhat trivial and probably not that important. A medium effect size ($d \geq 0.4 \leq 0.6$) may have importance. A large effect size ($d \geq 0.8$) is probably important. For a detailed discussion the reader should see Cohen.[5] There are several different web sites that automatically calculate Cohen's d.

So when you go to see your friendly biostatistician and they want to know what effect size you think is important, you can always say, "I think a Cohen's 0.7 might be reasonable."

WHAT IS A Z SCORE?

Sometimes statisticians will talk about using a standardized score or a z score. What the z score does is assign a value in terms of standard deviations to any data point. This value can be either positive (if it is to the right of the mean) or negative (if it is to the left of the mean). What the z score does is convert a raw score into one that uses standard deviation units. It is very easy to compute. One simply takes the data point and subtracts the group mean from that data point. You then divide the quotient by the SD.

$$z \text{ score} = \frac{\text{actual data point} - \text{the MEAN of the data set}}{\text{standard deviation}}$$

A z score distribution has a range of -3.99 to $+3.99$ and a z score of $+1$ means 1 SD above the mean and a z score of -1 means 1 SD below the mean. What this z score actually corresponds to is a point on the bell-shaped curve (normal distribution). Think about it as if you were riding down the bell-shaped curve and stopped along the way. You then dropped a straight line down to the X-axis you would get a value that goes from 0 to 3.99. That is the z score. If your data set is <u>not normally distributed</u> then you <u>cannot use a z score</u>.

When would you actually use a z score? Let us say that you wanted to compare the learning scores of your rats in a Morris water maze with that of your friend at another university. The means and standard deviation of the learning will no doubt be different because you both have differences in your mazes. You can make the comparison by changing all scores to some type of standard score which will have a common mean and common standard deviation.

DEGREES OF FREEDOM

Every time you run any statistic and want to determine the significance, it is essential that you have the **degrees of freedom** (df or DF). Every single statistical table requires this value and so it must be something really important. Actually it is very important and directly associated with the sample size. It is a type of mathematical restriction that is used when estimating the outcome of a statistic. What it represents is the total number of data values in your study minus some restriction. Let us just say that I gave you a number like 23, and told you there are a total of five values that add up to that number. I then told you that four of those numbers were 4, 8, 6, and 2. You would then know that the fifth number was 3. You would have no freedom really to pick the fifth number because I told you the total and already picked four values. If I only gave you three numbers (e.g., 4, 6, 2), then you would have two choices to make. You would have 2 degrees of freedom and be able to come up with a number of different combinations for the missing numbers:

1 and 10   2 and 9   3 and 8   4 and 7   5 and 6

If I only gave you one number, for example, 6, you would have even more possibilities. However, you have to first know what the total is supposed to be. In statistics, we know what the total is supposed to be and we know one of the numbers. So in the case where five numbers add up to 23 and one of the numbers is 6, we have four other numbers to be chosen, thus 4 degrees of freedom.

WHY n−1?

In many statistical equations there is a term, usually in the denominator, that is n−1. This is usually closely related to the degrees of freedom. When we have a data set, which is actually a sample, there are many different possibilities that these numbers could be. Let us just say that you had a defined population of 200 subjects and you needed to get a sample of 10 subjects from this defined population. Anyone of the 200 could be chosen, plus anyone of the remaining 199 is chosen next, and then any of the remaining 197, and so forth. There are over 1800 different samples you could draw. If the population is not well defined and consequently extremely large, then the number is absolutely huge. The bottom line is that you have only obtained a sample and that sample most probably does not accurately estimate the population it was drawn from, despite your best efforts to make it totally random. There will be some error involved in the estimation. To compensate for this error, the "n" in formulas is reduced by 1. The effect of this is actually to increase the variance, hopefully making your estimate more accurate.

SUMMARY

- The type of data you collect dictates what type of statistic can be used.
- Use the measures of central tendency that accurately describes your data.
- When describing the variance in a small sample use the standard deviation.
- It is important to have a large effect size for biological significance.

References

1. Braak H, Braak E. Neuropathological stageing of Alzheimer-related changes. *Acta Neuropath.* 1991;82:239–259.
2. Huck SW. *Statistical Misconceptions.* New York: Psychology Press; 2009.
3. Cohen JA. Power primer. *Psych Bull.* 1992;112:155–159.
4. Hedges LV, Olkin I. *Statistical Methods for Meta-analysis.* Orlando: Academic Press; 1985.
5. Cohen J. *Statistical Power Analysis for the Behavioral Sciences.* Hillsdale, NJ: Lawrence Erlbaum; 1988.

4

Graphing Data

Stephen W. Scheff

**University of Kentucky Sanders—Brown Center on Aging,
Lexington, KY, USA**

There is nothing more difficult to take in hand, more perilous to conduct, or more uncertain in its success, than to take the lead in the introduction of a new order of things ... because the innovator has for enemies all those who have done well under the old conditions... **Niccolo Machiavelli**

HOW TO GRAPH DATA

One of the more important aspects of statistical analysis is the representation of the results in a pictorial manner—graphing of data. However, in most cases (there are good exceptions), this should be done **after** statistical analysis has been carried out! Why??? Because, if you look at the graphed data, you will be biased in terms of what statistical groups you think should be compared. You will also be biased in terms of what you "believe" should be the statistical outcome of the analysis, and perhaps subconsciously use statistical techniques to fulfill that belief. Surely, you will want to use a graphing style that accentuates the differences you believe are the most important. Instead, at the time of formulating the research hypothesis and planning out the experimental design, before any data are collected, decide how the results should be displayed in a graph. I can hear you saying already, "But I'm not sure how the data are going to look before I collect them. Once I actually see the data I can get a better 'feel' for how they should be presented." Often the graphing style is somewhat "dictated by the field" and the existing literature. If you have researched your area of interest thoroughly (i.e., read as many papers as possible), you should have a fairly good idea of what type of data your research will generate. Perhaps you have run a pilot study first and have a clear idea of the type of data. One does not always have to be a lemming and follow the pack when presenting research results. Remember, researchers do need to be cautious not to misuse (maybe abuse) a graphing technique.

The most common graph type is the bar graph or column graph as it is sometimes called. This type of graph can be used to represent many different types of statistical results. For example, it can be used to show a frequency histogram.

In a recent experiment, an investigator chose to investigate neuron size in sections of cortex. He programmed his image analysis system to identify neurons in a defined counting frame. The system then evaluated the somal diameter (μm) of the neurons and gave the results seen in Table 4.1.

The range of scores appears to be from 7 to 53 μm and one can make a frequency histogram based on some prior information about neuron size or let a statistical program do it automatically. Very small neurons could be those with somal diameter sizes 7−14 μm and very large neurons from 45 to 53 μm.

Figure 4.1 shows a possible frequency distribution graph of the data in Table 4.1. Each column shows the number of neurons within the arbitrary

TABLE 4.1   Neuronal Somal Size (μm) Cortex Area 23B

49	49	25	46	47
32	49	26	22	49
36	35	28	18	41
32	27	19	10	15
33	38	43	20	29
22	47	29	33	31
18	24	53	45	30
47	28	52	34	29
29	37	36	31	8
42	18	28	17	21
33	14	37	25	19
14	20	35	47	43
9	50	7	12	43
33	43	32	21	14
29	14	39	40	15
43	33	32	44	8
30	34	32	34	15
31	49	21	16	14
11	13	52	7	7
27	38	38	45	32
10	8	20	15	51
12	42	23	21	31
35	14	38	45	33
37	42	38	45	52
7	43	44	26	30
37	16	23	50	49
40	18	17	27	34
9	36	41	46	48
29	27	25	28	32

Distribution of Cortical Neurons

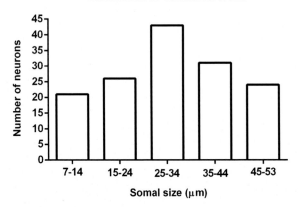

FIGURE 4.1   This is a frequency distribution of cortical neurons with a diameter range from 7 to 53 μm. The grouping of diameter size was arbitrary defined by the investigator. It shows a very normal distribution.

size category shown. How do you determine the range for each category? This is really a matter of preference or may be dictated by the literature. Most statisticians tell you to use what makes the most sense (whatever that means).

From Figure 4.1 it is easy to see that the most frequent neuronal size is in the range of 25–34 μm and that the neuronal size histogram follows a normal distribution (Chapter 3). This graph has a couple of important features: (1) it is simple and easy to interpret, (2) the axes are clearly labeled, (3) the Y-axis is calibrated sufficiently so that each column is easy to interpret, (4) each column is labeled, (5) each column has the same visual width, and (6) there is only a single idea communicated in the graph. Why is the frequency histogram better than showing all of the actual data points?

As can be seen in the scatter plot in Figure 4.2, while this graph shows the same information as the frequency histogram, it is not as easy to interpret. It is unclear how many more neurons fall into the 25–34 μm group compared to the 15–24 μm category. It is also difficult to visualize the normal distribution of neurons for this sample of the cortex.

A few quick words about graphs in general. The independent variable is plotted on the X-axis and the dependent variable is plotted on the Y-axis. For example, if you are plotting the cell firing rates as a function of four different drugs (A, B, C, D), the graph would be plotted like Figure 4.3. Also the size of the graph should be such that the Y-axis height is about 70% of the X-axis length, although this can vary if there are very few or very many independent groups.

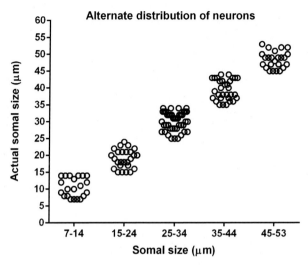

FIGURE 4.2   Scatterplot of the same data shown in Figure 4.1 using the same arbitrary group sizing.

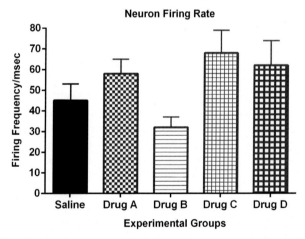

FIGURE 4.3   This graph shows a well-labeled vertical Y-axis and horizontal X-axis. Each of the groups can be easily distinguished by different shading. The error bars showing the variance are clearly indicated.

Nominal Data

This type of categorical data is normally either presented in a table or in a column type graph. One simply counts the number of neuron occurrences in each category and expresses the total as a number or even as a percentage of the total number of occurrences. In Figure 4.4, identifying

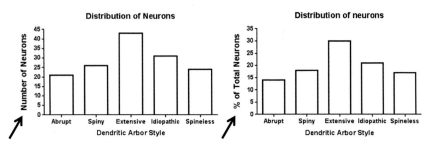

FIGURE 4.4   The graph on the left shows a distribution of the actual counts of neurons classified according to different dendritic arbor. One can get an estimate of the total number of neurons evaluated from this graph. These same data were transformed as a percent of total neurons evaluated. Note that the distribution did not change but the labeling of the Y-axis did. Note the difference in Y-axis labeling indicated by the arrows.

the type of neurons in the sample of the cortex might be based on the dendritic arbor instead of the somal size, but still represented as a frequency distribution.

It could also be plotted as a proportion of the total number of neurons evaluated. In this example there were 145 neurons identified and classified according to dendritic arbor. The "Abrupt" dendritic arbor neurons represented 14% of the sample while the neurons with "Spineless" dendrites represented 17%.

Here is another example of plotting nominal data. You might be interested in determining whether or not the quadrant where the submerged platform is in your Morris water maze (MWM) has any effect on the rat's ability to learn the task. A quadrant bias could have serious consequences on the interpretation of the behavioral data. A total of 40 rats are tested in the MWM with 10 subjects representing each of the four different quadrants where the submerged platform is located. Each group is trained for 5 days with the platform submerged in their representative quadrant. Here are the results of the testing (Figure 4.5). It appears that when the platform is submerged in quadrant I, more subjects have an escape latency less than 11 s on day 5. Perhaps there is some type of cue bias connected with the maze that makes it easier for the rats to locate a submerged platform in quadrant I.

Ordinal Data

This type of data has an important characteristic, in that the order in which the data is graphed is important. The neurons graphed as a function of somal size (Figure 4.6) is an example of this type of data and the frequency distribution is an appropriate graph. From this graph one can make a statement about the overall size of neurons in the sample of cortex,

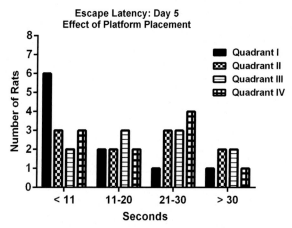

FIGURE 4.5   These water maze data were graphed as another example of nominal data. The grouping of the time spent in each quadrant was arbitrary defined by the investigator. A different arbitrary grouping could possibly result in an alternative interpretation of the data.

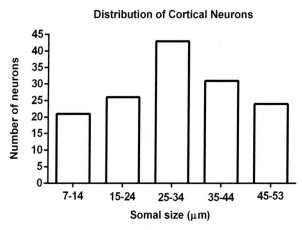

FIGURE 4.6   This is the frequency distribution of cortical neurons as in Figure 4.1 with a diameter range from 7 to 53 μm. The grouping of diameter size was arbitrarily defined by the investigator. Because it is a frequency distribution there are no error bars.

such as stating that the majority of neurons have a somal size greater than 24 μm, or that most of the neurons have a somal size less than 35 μm.

Continuous Data

There are multiple different ways to graphically display very simple interval and ratio data. However, not all methods give a complete picture of the data and subsequently do not lend themselves to full interpretation.

Consider an experiment involving a drug that you believe reduces lesion volume following trauma. There are three different doses of the drug Minimizerit® plus a vehicle (Saline) control group. Each group has 10 subjects and all subjects were evaluated at the same time post injury. Table 4.2 shows the data and Figure 4.7 shows a graph of the mean lesion volume of each group.

TABLE 4.2   Minimizerit and Lesion Volume

Saline	5 mg/kg	10 mg/kg	20 mg/kg
62	47	43	40
47	43	37	21
56	34	33	37
44	35	37	27
58	37	29	38
37	47	36	18
57	31	42	42
37	39	35	19
63	48	24	19
54	33	22	32

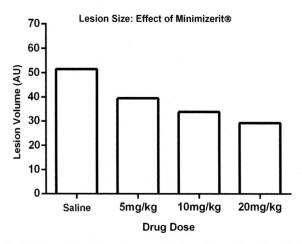

FIGURE 4.7   Basic representation of group data from Table 4.2 on the effects of Minimizerit on lesion size. Without information concerning the variance within each group the interpretation of the data is limited.

While the graph certainly represents the data of each group, one has limited information about the actual differences that the drug has on the lesion volume, other than the fact that the drug appears to have reduced lesion volume. What would be extremely helpful is to show the variance within each group. Very often such data are graphed as the mean of each group and the standard error of the mean (SEM). What this type of graph portrays is the idea that there is very little variance within each of the groups and there is virtually no overlap of values when comparing one group to another (Figure 4.8).

However, if one carefully looks at the actual data it is quite clear that there is considerable overlap between groups. Some animals in the 20 mg/kg group have values within the range of the Saline control group. Obviously, this graph of means ± SEM is very misleading, and gives the impression that the effect of the drug is extremely predictable.

A reason that individuals often state for displaying data using the SEM is that it makes the results "look better." The SEM is an inappropriate statistic of the variance in this experiment. (See Chapter 3 for a discussion of standard deviation and standard error of the mean.) A more correct display of the variance would be to use the standard deviation (SD) (Figure 4.9).

Since ± 1 standard deviation represents roughly 68% of all the scores for a given group, it is now possible to begin to get an idea of how much variance there is in each group compared to each of the other groups. If you know that 2 standard deviations represent approximately 95% of all the scores in a group, it becomes clear that there is considerable overlap

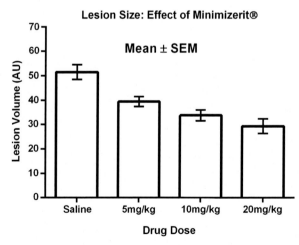

FIGURE 4.8   This is the same graph as in Figure 4.7 with the exception that it shows error bars as the standard error of the mean (SEM).

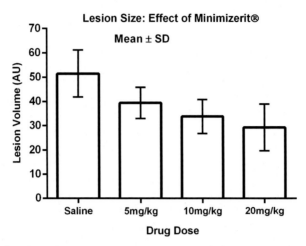

FIGURE 4.9   This is the same graph as in Figure 4.8 with the exception that it shows error bars as the standard deviation (SD).

between the different groups. Consequently, if another investigator were to run this experiment using only the 20 mg/kg dose, they might expect to find values in the range of 10–48.5 (mean ± 2 SD) if the data are normally distributed. From the SEM data the range would be 23.2 to 35.4 and clearly the scores were outside of the range.

BOX AND WHISKER PLOTS

Another common way to display these data is using what is called a **box and whisker plot**. Figures 4.10 is an example of the above data plotted with boxes and whiskers. It may look somewhat strange and a little difficult to interpret unless one knows how it is constructed.

Basically what each box shows is the first, second, and third quartile, which in itself sounds rather strange. The line in the middle of each box represents the median of the group data. The whiskers (T) represent "a type of range for the quartiles."

Here is how a **box and whisker plot** is generated and perhaps you can get a better feel for the interpretation (Figure 4.11). Let us consider the data for the Saline group in Table 4.2 and use all 10 subjects.

1. 62, 47, 56, 44, 58, 37, 57, 37, 63, 54
2. Arrange the data in ascending order 37, 37, 44, 47, 54, 56, 57, 58, 62, 63
3. Find the **median** (middle most score) 37, 37, 44, 47, **54, 56,** 57, 58, 62, 63

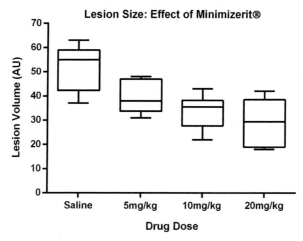

FIGURE 4.10   Typical box and whisker plot using the data in Table 4.2.

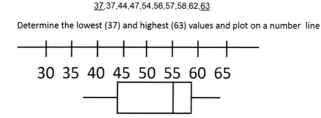

37,37,44,47,54,56,57,58,62,63

Determine the lowest (37) and highest (63) values and plot on a number line

Draw vertical line through median

Draw box around 1st & 3rd quartile

Mark each of the following on the number line
Median (55)
First quartile (44)
Third quartile (58)
Lowest value (37)
Highest value (63)

Draw line (whiskers) from each
end of box to the lowest and
highest values

FIGURE 4.11   A further explanation of how a box and whisker plot is constructed.

    a. because there is an even number, take the score half way between scores 5 and 6 = 55)

    b. this is called the **2nd quartile**

4. Find the median of the lower half 37, 37, **44**, 47, 54 = 44

    a. This is called the **1st quartile**

5. Find the median of the upper half 56, 57, **58**, 62, 63 = 58

    a. This is called the **3rd quartile**

6. Determine the lowest (37) and highest (63) values and plot all of these on a number line.

Sometimes there is a value that is not even close to the other values and is considered to be an "extreme outlier." Just to demonstrate the point the lowest point could be a 12 instead of 37. In this case the ascending order of the values would be:

12, 37, 44, 47, 54, 56, 57, 58, 62, 63

The box and whisker plot would look almost identical with the exception that the extreme outlier (12 indicated by arrow) is marked with an asterisk as shown in Figure 4.12.

The box and whiskers can be plotted in a horizontal fashion or in a vertical manner as above for the entire data set with the Y-axis and X-axis marked appropriately.

There are several very positive aspects about this type of presentation of the data.

1. If the data set is very large, it provides a very useful display of the data distribution that is easy to understand.
2. If the median is not located in the middle of the box, then you know that the distribution is skewed to one end.
3. Outliers are easy to recognize and these values do not necessarily bias the graphic representation of the data.
4. It is an excellent way to present data that do not have a normal distribution.

There is also a major drawback to box and whisker plots.

1. Often they are generated by a computer program that may use a different set of rules for constructing the box and the whiskers.
   a. The line may represent the mean instead of the median.
   b. Whiskers may be plotted to the highest and lowest values and outliers not shown. In this case the reader may believe that there are many more points dispersed throughout the range of that whisker. It would be imperative then to understand how many data points are used to make up the plot.
2. One really does not have any idea of how many data points are represented in such a plot.

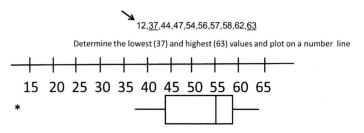

FIGURE 4.12   Note how the asterisk is used to indicate an outlier (12) in the data set.

The box and whisker plot is becoming more common in the clinical neuroscience literature and the ability to understand the meaningfulness of this display of the data takes some practice. The major keys to interpretation are to remember that:

1. The ends of the box represent the medians of the top and lower halves of the data;
2. Unless stated, the line in the middle of the box represents the **median** and not the mean;
3. The ends of the whiskers, unless otherwise stated, represent the highest and lowest scores.

Here are three different groups of values to demonstrate differences in the look of a box and whisker plot (Table 4.3). The Mean and SD are provided also.

Group 1 has a pretty even distribution of scores (not skewed) and hence the median line is in the middle of the box (Figure 4.13). Since the whiskers are approximately the same length both above and below the box, the data appear to be spread out evenly over the range with no apparent outliers.

Group 2 has almost the same median score as the first group with the exception that it sits closer to the bottom of the box. This indicates that the data are skewed toward the lower end of scores. Whenever the middle line (the median) is located closer to one end of the box, it means that half of the data are in that smaller portion and thus skewed. The very small bottom whisker means that there is a very tight range of scores in the lower half of the data and that the high values have a much greater variance.

Group 3 has almost the opposite data distribution from Group 2. As indicated by the size of the box and also the whiskers, the range of scores is much larger than the other two groups. The scores are also skewed toward the higher scores. If you look at the third column of numbers in Table 4.3, it is easy to see that the range of scores is from 53 to 62 for that half of the data. It is also easy to see from this plot that there is considerable overlap with the other two groups (Figure 4.13).

Quickly learning to read a box and whisker plot takes time and practice. It does provide much greater information than a standard column (bar) graph with the standard deviation plotted.

SCATTER PLOTS

If the number of subjects within a group is 25 or less, then the graphing option that provides the greatest information is one that displays each data point used in the analysis. Nothing is left hidden to the reader. Many investigators "shy" away from this type of graph because it reveals all of the variance. Knowing the variance is at the heart of understanding the

TABLE 4.3   Box and Whisker Plot Data
of Luminescence

Group 1	Group 2	Group 3
18	27	19
20	27	26
21	29	26
22	30	29
23	30	30
23	30	34
23	31	40
26	32	46
26	32	50
29	33	53
30	35	53
31	36	56
32	39	57
33	39	58
35	43	58
35	44	59
36	46	60
38	52	61
39	55	61
41	58	62
$\overline{X}$ 29.1	$\overline{X}$ 37.4	$\overline{X}$ 45.4
±6.9	±9.4	±15.9

meaningfulness of the data. Compare a graph that represents the Mini-mizerit data showing all the data points with the bar graph and the box and whisker plot (Figure 4.14).

There is absolutely no confusion about the dispersion of the data in each group when displayed as a scatterplot. In this particular scatter plot example, the line represents the median. To help in understanding why the SEM is an inappropriate measure of the variance and the SD is more appropriate, consider these two plots with the actual data from the scatterplot superimposed (Figure 4.15). Almost all of the data points are

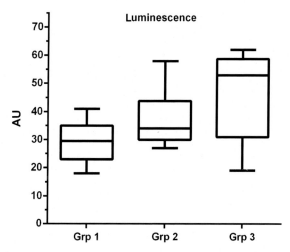

FIGURE 4.13 Box and whisker plot of the data in Table 4.3. Note the change in the box size for each group.

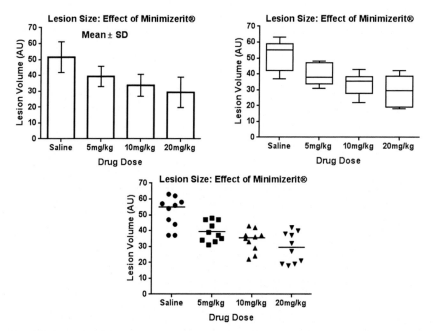

FIGURE 4.14 Composite of three different graphs representing the same data in Table 4.2. Which of these graphs conveys the most information about the data and is the easiest to interpret?

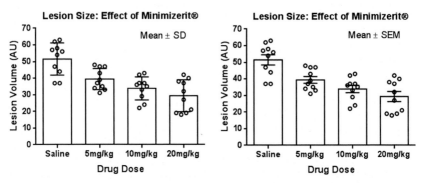

FIGURE 4.15   Graphs of the data in Table 4.2 using a commercially available graphics program. Note that the program overlays error bars and columns on top of the scatter plots. Sometimes commercially available programs use the SEM as their default setting as shown in the graph on the right. It is important to investigate default settings.

contained within the SD in the graph on the right, while almost all of the data points are above or below the SE lines for the graph on the right. Plotting the SEM hides the variance. Let me stress that <u>it is **inappropriate** to plot a column graph showing the SD along with the actual data points such as in</u> Figure 4.15. It is done here merely to make a point about using the SD instead of the SEM. If you are showing the scatterplot then the reader has a complete view of the variance and thus the columns and the error bars are superfluous.

ALTERNATIVE GRAPHING PROCEDURES

Sometimes researchers obtain data that show an effect that they simply wish were not there, or they get data that just do not appear to support their idea. Often these findings are simply omitted from a manuscript with the hopes that the reviewer will not ask for it. The pictorial display of those data can be manipulated by simply changing the range of the Y-axis. Here is an example of both of these manipulations of which one should be very conscious.

The experimenter is using a very expensive compound to determine if the novel drug can reduce lesion volume following a particular perturbation. After carefully analyzing the results he finds that the novel compound does reduce injury volume, but there is also a dose–response. The highest drug dose shows the greatest reduction. However, because the drug is so expensive, he would rather simply state that while the drug does significantly reduce lesion volume, there is not a significant difference between drug groups. Consequently he will use the 5 mg/kg dose in subsequent experiments.

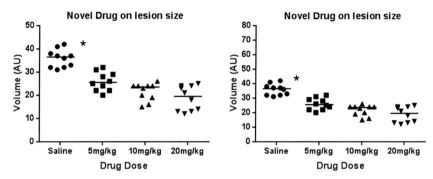

FIGURE 4.16 These two graphs show the exact same data (Table 4.2) but the one on the right was configured to downplay the differences between group means and also the variance.

By simply changing the range of the Y-axis (Figure 4.16), the visual appearance of the dose–response effect is minimized. While the data are exactly the same in both graphs, the intent is to deceive the viewer.

In another experiment the researcher was disappointed in the magnitude of the effect of his novel compound because the effect appeared to be minimal. Unless he could demonstrate something sensational the granting agency probably will not be receptive to his idea. The initial graphing of the results is shown on the left in Figure 4.17 and the "enhanced" version is shown on the right. The results are now quite impressive. If a reviewer looks at the graphs very quickly, the impression is that the drug has a significant biological effect. It now appears that the novel compound has almost eliminated the lesion volume with the highest dose of the drug. While there really is not any rule against doing this, when reviewing graphs in published papers one should pay very close attention to the scale of the Y-axis.

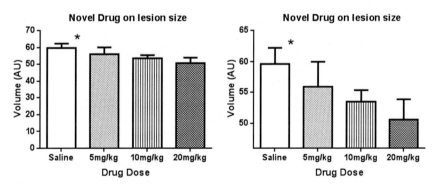

FIGURE 4.17 These two graphs plot the exact same data (Table 4.2) but the one on the right has been modified to accentuate possible differences between groups. Note that the range of the Y-axis is different between the two graphs.

INDICATING SIGNIFICANCE ON A GRAPH

There is really no convention regarding the showing of significance on a graph, although some journals have a few standards. The use of a single asterisk (*) is often used to signify significance at the 0.05 level and a double asterisk (**) signifies 0.01 and three asterisk (***) signifies 0.005 or greater significance. Depending on the experimental design and the ferociousness of the analysis, there can be quite a few comparisons shown on a single graph. The standard rule is to keep the graphical interpretation simple.

In Figure 4.18 each of the drug groups was compared to a single Saline group. Consequently the indicators of significance are simple and the figure legend associated with the graph would indicate what each asterisk indicated. It is understood that each drug group was not compared to any other drug group.

In Figure 4.19 there are multiple different comparisons. The asterisk is still used to indicate comparison to the Saline group and the lines then indicate the other multiple comparisons. A different indicator is used to show the comparisons that are not with the Saline group. In this case, the figure legend would indicate what a single, double, or triple hashtag indicates. It is a good idea to refrain from using different small case letters as indicators of significance in graphs and to try and limit how many different comparisons are shown. Some statisticians believe that only the significance of the main effect should be indicated and other comparisons described in the text or a table.

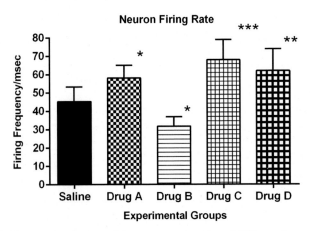

FIGURE 4.18   Graph of neuron firing rate as a function of different drugs. Bars represent group means ± SD. *$p < 0.05$; **$p < 0.01$; ***$p < 0.005$ compared to the Saline-treated group.

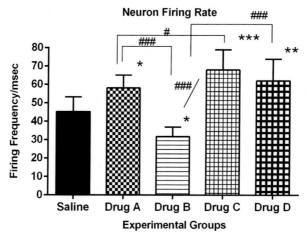

FIGURE 4.19   This is the exact same data as shown in Figure 4.18 with all of the different group comparisons shown. It is important to indicate which comparisons are associated with which symbol. It is a good practice to use lines to indicate the comparisons. Bars represent group means ± SD. $*p < 0.05$; $**p < 0.01$; $***p < 0.005$ compared to the Saline-treated group. $\#p < 0.05$; $\#\#\#p < 0.001$ with comparisons as indicated.

SUMMARY

- Keep graphs simple because they are important in analysis presentation.
- Make sure that graphs are properly labeled.
- Always use a graph that is appropriate for the type of data collected.
- Showing the variance in an experiment is extremely important.
- Use a graph type that clearly describes the variance.

Correlation and Regression

Stephen W. Scheff

University of Kentucky Sanders—Brown Center on Aging,
Lexington, KY, USA

*Correlation doesn't imply causation, but it does waggle its eyebrows
suggestively and gesture furtively while mouthing — look over here.* **Anonymous**

CORRELATION

A simple correlation is a measure of the relationship between two
variables. A **correlation coefficient** is a numerical measurement of the
strength of that relationship. It has a range from -1.00 to $+1.00$, so if you

Fundamental Statistical Principles for the Neurobiologist
http://dx.doi.org/10.1016/B978-0-12-804753-8.00005-1

see a correlation of 1.37 you know there is a problem. Most correlations describe the relationship between two variables and can also be called a **bivariate correlation**. A correlation differs from a **regression** in one important aspect. A correlation makes no *a priori* assumption (Chapter 6) about the relationship between the variables and can be considered a test of interdependence. A regression makes the assumption that there is a cause−effect relationship between the variables. The regression actually attempts to describe the dependence of one variable on another. This actually is very hard to prove.

A correlation value of zero (0) means that there is absolutely no linear relationship between the two variables.

- Statisticians would say that the data are "uncorrelated."
- A correlation of zero rarely exists, except in cases where one of the variables under consideration never changes.

A correlation value of +1.00 or −1.00 means that there is a perfect linear relationship between the two variables.

- A positive value indicates an "uphill" or positive relationship such that as the numeric value of one variable gets larger so does the numeric value of the other variable.
- Statisticians would say the data are "positively correlated."
- A negative value indicates a "downhill" or negative relationship such that as the numeric value of one variable gets larger the numeric value of the other variable gets smaller.
- Statisticians would say the data are "negatively correlated" or "inversely correlated."
- The term **anticorrelation** has also been used but is not very popular.

A correlation of −0.68 is stronger than a correlation of +0.47 since the negative or positive sign only indicates the direction of the relationship. It is the absolute value of the coefficient that indicates the strength of the relationship. It is a mistake to believe that a negative correlation is "bad" and only a positive correlation is "good." As stated above, the sign before the correlation only indicates the direction of the relationship between the two variables and the nature of this relationship depends on the nature of the experiment itself. For example, in a case where one was measuring the size of a brain tumor and the individual's ability to perform a specific cognitive test, an increase in the size of the tumor could negatively correlate with the individual's cognitive ability (Figure 5.1). As the tumor became smaller the cognitive ability might become better.

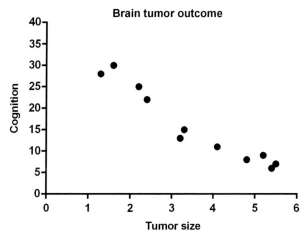

FIGURE 5.1 This is an example of a negative correlation showing that as tumor size increases the cognitive score decreases.

PEARSON'S PRODUCT–MOMENT CORRELATION COEFFICIENT

This is a very commonly used parametric statistic that measures the linear relationship between two ordinal/interval variables. Almost every commercially available statistical package includes the ability to perform this statistical analysis. These variables must be continuous in nature. It was developed by Karl Pearson in 1895 and sometimes called Pearson's r, Pearson's rho, Pearson's correlation coefficient, PPMCC, or simply **P** (not to be confused with the significance p value).

In a recent experiment, the investigator was interested in determining whether or not there was a relationship between the injury volume that she observed following a controlled cortical impact and the weight of the animal. She randomly chose the injury results from several different experiments and then plotted the results and the rodent weight at the time of the injury. All of the animals received the same magnitude of injury using the same experimental device. One of the important conditions before analyzing the data with the Pearson's rho is to determine if it is linear. This is one of those cases where it is wise to graph the data before running the analysis. To the investigators dismay, at first glance there does not appear to be a major difference, but by changing the dimensions of the Y-axis it is possible to gain a better understanding of the possible relationship before attempting to analyze the data.

There is no intent here on deceiving anyone but rather to better understand the linear relationship of the data. Sometimes one must plot the data over a narrow range to see the true linear relationship such as in Figure 5.2.

The graph on the right in Figure 5.2 appears to show a linear relationship between the weight of the subject and the injury volume. The two variables are interval in nature and have a continuous range. These are important criteria for using the Pearson's rho. The rho for this set of data is $r = -0.611$, which is significant at the $p < 0.02$ level. For a correlation, it does not matter which variable is plotted on the X-axis and which is plotted on the Y-axis since a cause–effect relationship is not under consideration, even though there may be one. To find the significance of the correlation coefficient one calculates a t value and looks up the significance in a Student t-distribution table. The Student t-distribution is easily located on the web and there are many different examples. Since almost every commercially available statistical package does this automatically it is not worthwhile to show that equation here. This t-statistic is rather unique in that to calculate the **degrees of freedom** one uses the total number of pairs of scores and subtracts 2. The concept of degrees of freedom is extremely important in statistics and also very important when determining the significance of a particular operation (see Chapter 3).

For the above experiment the t value $= 2.78$ and since there were 15 pairs of scores the df $= 13$. In the Student t-distribution table for a two-tailed test a value of 2.65 is necessary for significance at the 0.02 alpha level and 3.012 for the 0.01 alpha level. Since the t value in this experiment is in between these two values we report $p < 0.02$ or $p < 0.05$.

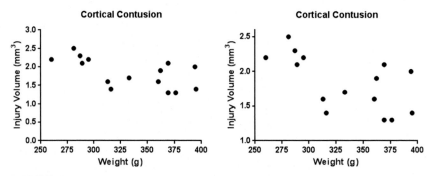

FIGURE 5.2   When plotting a correlation it is often helpful to use a limited range on the Y-axis, as shown in the graph on the right, to gain a greater appreciation of the relationship between the variables.

What Does the Pearson's Rho Actually Test?

This statistical test actually evaluates the probability that the relationship observed between the two variables could have occurred by chance. In other words, the H_0 would be that there is no relationship between the two variables. While it is true that a correlation coefficient does not necessarily mean that there is a cause–effect relationship, it can certainly imply that such a relationship exists. A correlation coefficient indicates the strength of a linear relationship between the X variable and the Y variable.

Often one will see the results of a correlation presented as r^2. While most researchers will call this r squared, the statistician will say this is the **coefficient of determination**. One simply squares the Pearson's rho (r) value. What this actually does is describe the proportion of variance that the two variables under consideration share. In this case $r^2 = 0.373$ and no sign is placed in front of this value. This means that 37.3% of the variance in the data set is shared by the two variables. In other words, the amount of variability (in this case 37.3%) that you see in one variable, such as Y, can be explained by its relationship to the variable X. Reporting the r^2 value on a graph of a correlation coefficient is really <u>not correct</u> and one should report only the r value. The r^2 value is really only appropriate when graphing a regression (see below).

How Many Subjects are Required to Perform a Correlation?

This is actually a difficult question to answer because it depends on the level of significance (alpha level), the effect size that the investigator is interested in testing, and the power of the test. There is not any predicated number of subjects and often you will see it applied to only six or seven subjects. Many statisticians would say that you should have at least an n = 20 to begin to have a reliable correlation and yet others would say you need at least an n = 25 or n = 30. Many statistical programs have a calculator that will determine these values for a given power and the reader is advised to read the section on power in this book. As an example, let us say that in a preliminary set of data (n = 4) you got a correlation coefficient of 0.312 and want to know what sample size is necessary to get significance at p < 0.05 with 80% power. It would require approximately 80 subjects. If the power was increased to 85%, then approximately 90 subjects would be required. Alternatively, if the correlation coefficient in your preliminary study was greater, such as 0.450, then the number of subjects necessary to get significance at p < 0.05

with 80% power would be only 37. Remember, to get significance with a small effect size, a much greater sample size is required.

Reporting a Correlation Coefficient in a Manuscript

When reporting the results of the relationship of X and Y data using a correlation analysis one needs to give the r value, the p value, and the number of subjects used in the analysis, such as $r_{(15)} = -0.611$, $p < 0.02$. This is often shown in the body of a graph of the data. When the individual points are shown it is not necessary to give the sample size. However, it is incorrect to plot a regression line when plotting a correlation (see Figure 5.3). A regression line (described later) assumes a cause—effect relationship and a correlation does not necessarily demonstrate this even though it may be the case. The point of a regression line is to be able to predict response Y when given variable X. The relationship between two sets of data is actually the scatterplot. The regression line is actually an adornment used to predict mathematically the relationship of the two variables.

A text format for reporting a Pearson's correlation might be: A Pearson's correlation coefficient was carried out between the body weight and the injury volume in 15 adult rats following a cortical contusion. There was a significant negative correlation between body weight and injury volume ($r = 0.611$, $n = 15$, $p < 0.02$).

A final word of caution when applying the Pearson's correlation coefficient—it is very sensitive to points that are outside the range of all the other points and these can be considered "outliers." As seen in

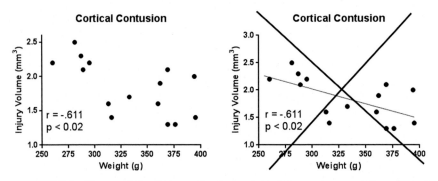

FIGURE 5.3   The correlation between the change in the injury volume and the weight of the rat is correctly graphed on the left. However, the graph on the right also includes a regression line and this is not appropriate since a correlation does not mean a cause—effect relationship and a regression line indicates this.

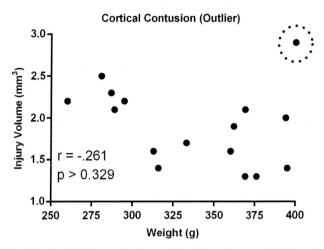

FIGURE 5.4   The data in this correlation are identical to those used in Figure 5.2 with the addition of one extra data point indicated by the broken circle. This single point has a significant influence on the correlation as indicated by the reduced r value and increased p value. These "outlier" data points should be scrutinized very carefully for accuracy.

Figure 5.4, a single data point (indicated by a dotted circle) totally changes the analysis. In this particular case, a single outlier has enough influence that it lowered the correlation from 0.611 to 0.261, which is no longer significant at the $p < 0.05$ level.

If there is an obvious outlier, it is important to carefully check the accuracy of that data point and whether or not its value was correctly entered into the data set. It sometimes happens that, due to pure chance, a subject "performs" differently. If the data point appears to have high suspicion associated with it then you might have justification for removing it (see Chapter 9). On the contrary, if it appears to be a valid data point, then one can analyze the data set with this point included and also with it excluded and discuss this difference.

The most accurate correlations are obtained when there is a wide range of values for both the X and the Y variables. In the above example, if the animal weights were restricted to be between 300 and 400 g then the r value would change dramatically and the correlation would not be significant (see Figure 5.5).

SPEARMAN'S RANK COEFFICIENT AND KENDALL'S TAU

Besides the Pearson's rho statistic, there are two other correlation methods that are often seen in the scientific literature. Both of these are discussed in Chapter 8. The **Spearman's rho** correlation statistic is used

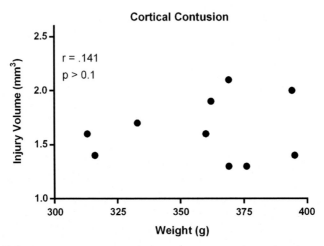

FIGURE 5.5   This correlation plot shows only a subset of the data from Figure 5.2. Because there are so few data points graphed the range of scores is limited and the result is a loss of significance. The greater the number of data points the more realistic will the correlation reflect the population from which it was drawn.

with nonparametric data analysis and was actually derived from the Pearson's coefficient. It is reported quite often in neuroscience literature. Essentially this statistic shows the relationship between two sets of ranks of data instead of two sets of quantitative data. Like the Pearson's rho, the values for the Spearman's rho go from -1.00 to $+1.00$. One of the unique features of the Spearman's rho is that it can be used when one of the variables is quantitative and the other is ordinal. **Kendall's rank-order coefficient** or **Kendall's tau** is a somewhat more complicated method that is not really based on the product–moment concept like Spearman's and Pearson's. It is not used as often as Spearman's. It actually looks at the distributions of the two sets of data. For a more detailed discussion of the nonparametric procedures see Siegel and Castellan.[1]

REGRESSION (LEAST SQUARES METHOD)

In many experiments, the investigator would like to make future predictions based on the outcome of a particular study. For example, does the increase in stimulation of the fimbria result in an increase in neuron firing rate of the hippocampal granule cells and if so can I predict that increase? If one has enough data he or she can plot a regression line, which describes the dependence of a particular variable (firing rate) on another variable (fimbria stimulation). This assumes that there is a one-way cause–effect relationship. If the relationship is strong then the experimenter can predict the outcome with a certain amount of confidence.

In a recent experiment, a graduate student wanted to find out how many trials were required for a group of rats to learn a new X47 maze that she developed. There were 25 errors possible that a subject could make in the maze, and she set a learning criterion of 80% correct. Figure 5.6 shows a graph of the results of testing six animals repeatedly in the maze. Each point represents the mean number of errors for all six subjects.

This graph plots the number of trials on the X-axis and the dependent variable (Errors) on the Y-axis. There appears to be a strong relationship between the two variables and the number of trials in the maze appears to predict the number of errors. From this graph, the experimenter can "predict" that by trial 11, most of the animals have obtained criterion. In this case, there is a cause–effect relationship. For example, the straight line is called the **regression line** or **line of best fit**. Does this mean that if she ran another group of subjects in the maze they would also reach criterion on day 11? Not necessarily but these data could certainly be used as a guideline to base future experiments. A statistician might say that you might use the **least-sum-square method** to find the best fitting straight line. Almost every statistical software package does this automatically. Briefly, what the computer program does is look at what straight line can be applied to the data points and then minimizes the distance between each individual point and that line. While there are formulas for doing this, it is quite time-consuming and even after you have the equation you then have to enter in values for the X variable and plot the subsequent Y value. The actual distance between any individual point and the regression line is called the **prediction error**. The vertical line that can be drawn between any point and the regression line is called the **residual**. Some points have positive residuals, if they are above the line and others

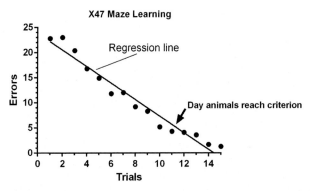

FIGURE 5.6   The data from the X47 maze learning study were plotted in a fashion similar to that of a correlation but there was definitely an effect of the number of trials in the maze that resulted in a decrease in the number of errors. Because there was hypothesized to be a cause–effect relationship, the investigator was able to show a regression line.

have negative residuals because they are below the line. If all the points actually fell on the line there would be no residuals because there would be no prediction errors. The question then becomes how accurate is the regression line at actually predicting a Y value when given an X value. This value is called the **standard error of estimate**. The standard error of estimate is the standard deviation of the prediction errors.

If the investigator is interested in establishing a cause—effect relationship then the design of the experiment should be one in which **one** of the variables can be manipulated. That variable is the independent variable and it is always plotted on the X-axis. In a recent study an investigator was interested in determining whether or not a change in the health of mitochondria changes the number of lipofuscin-containing neurons in the ventromedial nucleus of the hypothalamus in a cohort of Agouti rats. Different concentrations of a mitochondrial poison were injected into the ventromedial nucleus. At 4 days post injection, the nucleus was evaluated for lipofuscin-containing neurons from histological slides by individuals who were blinded with regard to the level of toxin. Because there were some technical issues during some of the injections, the n/group was not equal. The average number (cells/mm^3 × 10^2) of lipofuscin-positive cells was calculated (Table 5.1).

The findings were plotted and a linear regression line was applied to the data (Figure 5.7). There appears to be a very strong association between the increased toxin and the density of lipofuscin-containing cells in the hypothalamus. The line indicates that as the concentration of toxin increased so did the number of positive cells. Nevertheless, the line does not make it possible to actually predict the number of positive cells in the hypothalamus even if we know the toxin levels. The reason one cannot predict the number of cells, regardless of the overall significance of the association, is because the variability is too great at many of the concentrations, especially for the 18—25 pg/μl range.

The slope of the regression line can be tested for significance. Depending on the statistical software this will be given as either a t score or an F score, but both provide the same information. It is not necessary to provide these values when listing the significance, only the r^2 and the alpha level as shown in Figure 5.7.

Some journals insist that the confidence interval for the regression line also be displayed such as in Figure 5.8. This is because of the uncertainty in the estimate of the slope of the line. What this interval provides is a measure (usually 95%) of how much the line could deviate and still represent the population from which the sample was drawn. The lines representing the confidence interval always show the interval to be wider at the ends than in the middle because the regression line must be straight and go through a specific point on the graph defined by the means of the two variables.

TABLE 5.1   Changes in Lipofuscin-Positive Cells in the Hypothalamus

pg/μl	Cells/mm^3 × 10^2	pg/μl	Cells/mm^3 × 10^2
2	0.2	10	4.3
3	1.0	28	20.0
10	3.9	2	0.35
18	6.5	12	7.2
22	7.6	6	3.0
18	4.2	22	11.9
6	3.4	28	18.7
24	10.8	3	1.4
24	16.2	12	3.9
3	0.25	10	5.1
6	1.8	18	12.4
6	4.0	12	5.7
28	15.8	24	15.3
2	0.25	3	0.50
10	6.4	6	2.4
22	15.1	10	4.7
28	22.3	2	0.30
6	3.7	18	9.8
22	13.6	24	9.1
18	14.7	3	0.26
12	4.9	18	8.5
10	3.2		

Very often the statistical software will provide a **regression equation** that represents the regression line. In this example it is: $Y = 0.6488*X - 1.529$. What this equation is supposed to do is allow the investigator to provide a value for X and the equation will provide an appropriate estimate value for Y. This will only work if the value provided for X is reasonable (e.g., 15 pg/μl). Using a value such as 60 pg/μl is outside the realm of reality and will not result in useful information. The bottom line is, only use a number that makes biological sense. Often a biostatistician will talk about the **regression coefficient**, which is the rate

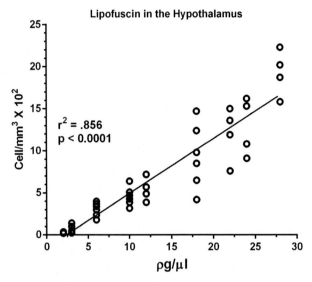

FIGURE 5.7   The hypothesis for this experiment centered on changes in mitochondrial function and the occurrence of lipofuscin in a select region of the hypothalamus. Because the experimenter was able to alter the level of mitochondrial poison, thus changing the mitochondrial function, the data could be analyzed and graphed as a regression experiment. In this case there was a very strong cause–effect relationship.

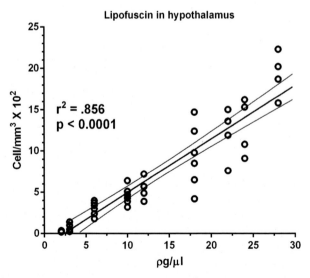

FIGURE 5.8   These are the same data as those in Figure 5.7 with the addition of lines showing the confidence interval. Several journals now require that all regression lines be graphed with the confidence interval.

of change of one variable as a function of changes in the other, or simply the slope of the regression line. It is totally unnecessary to place this regression equation on the graph.

With a simple linear regression, two variables are used with the intent of predicting the value of one variable from the other. When more than two variables are used to predict an outcome, one no longer has a simple linear regression but instead a **multiple regression**. There are many different ways of modeling biological data using multiple and **nonlinear regression**. When you are confronted with this type of situation it is best to talk with your friendly neighborhood biostatistician. Biostatisticians are experts in these types of analyses. You simply have to ask them for help in curve fitting your data.

SUMMARY

- A correlation simply describes the relationship between two variables.
- A regression describes a cause–effect relationship between two variables.
- It is inappropriate to plot a regression line on a correlation graph.
- A Pearson's correlation is a parametric statistic.
- A Spearman's rank correlation is a nonparametric statistic.

Reference

1. Siegel S, Castellan NJ. *Nonparametric Statistics for the Behavioral Sciences*. Boston: McGraw Hill; 1988.

One-Way Analysis of Variance

Stephen W. Scheff

**University of Kentucky Sanders-Brown Center on Aging,
Lexington, KY, USA**

*There are two kinds of statistics; the kind you look up and the kind you make
up.* **Rex Stout (Death of a Doxy)**

Fundamental Statistical Principles for the Neurobiologist
http://dx.doi.org/10.1016/B978-0-12-804753-8.00006-3

ANALYSIS OF VARIANCE

This statistic is one of the most abused in all of neuroscience primarily because it is so misunderstood and so incorrectly relied on to answer a specific question. In its simplest form, a one-way analysis of variance (ANOVA) is called a **t-test**. A t-test can be used to compare the difference between group means in an experimental design. What an ANOVA does is examine the amount of variance in the dependent variable and tries to determine from where that variance is coming. The question is really quite simple: Do the group means differ significantly because they truly come from different populations, or do the group means differ because of some random variation in the sample data? First we need to consider the t-test.

STUDENT'S t-TEST

This parametric statistic was first introduced by William Gosset in 1908, who used the pen name "Student." He was a chemist who was working for the Guinness brewery in Ireland and needed a quick way to monitor the quality of its beverages. This statistic allows the investigator to determine if two groups come from independent populations or are simply samples from the same population, with differences simply due to random sampling. It does this by comparing group means and the variance.

Unpaired t-Test: Testing the Means of Two Independent Samples

This is a routinely employed statistic that can be used to test the H_0 that two group means come from the same population. Most people would say it tests whether or not the two groups are statistically different. As an example, let us say that a researcher has developed a new drug, Smartamine®, that he believes will make rats perform better in a maze. He tests two groups of adult male rats and records the number of errors they make in learning to navigate the maze over a set number of trials (Table 6.1). One group is administered a placebo and the other his new drug Smartamine®.

Since the only relationship that these two groups have in common is that they both came from the same breeding colony, and were randomly assigned to either the placebo or the drug treatment group, they are considered to be independent samples and not paired. This is important because these are important conditions for using this type of t-test. In

TABLE 6.1 Smartamine® and T-Maze No. 1

Placebo	Smartamine®
47	38
50	45
52	42
36	29
47	39
39	37
45.2 ± 6.3	38.3 ± 5.4

addition, the values in each group need to be normally distributed and have standard deviations that are fairly close. If either of these latter two conditions is **not** met, then the investigator should use a nonparametric statistic such as the Mann–Whitney U test (Chapter 8). Virtually every commercially available statistical software package has the unpaired t-test as an option. For these data, the statistical software computes a t value $= 2.012$ with 10 degrees of freedom (df) and $p = 0.072$, which means that the H_o is not rejected. Thus, it looks like Smartamine® is not the next super drug for learning and memory. But perhaps this was only a pilot study and $p = 0.072$ is promising so why not repeat it with a larger n/group.

So the researcher decides to try the experiment again with 10 animals per group and records the number of errors they make in learning the maze. To make sure that the testing is performed in an unbiased way, the researcher asks a technician from another laboratory to administer the drug. After all of the testing is done, the drug group designation is revealed to analyze the results. However, there was a bit of confusion over which animals should be getting the drug and which the placebo. It turns out that more animals received the drug Smartamine® than the placebo (Table 6.2).

TABLE 6.2   Smartamine® and T-Maze No. 2

Placebo	Smartamine®
51	37
39	44
48	34
46	41
49	33
50	37
42	39
	39
	45
	35
	32
	42
	37
46.4 ± 4.4	38.1 ± 4.1

Is there a problem that the n/group is unequal (or as a statistician would say—"unbalanced")? Actually no, because the unpaired Student's *t*-test only requires that the two groups are independent and that variance between groups is not extremely different (homogeneity of variance which is covered later). For these data, the statistical software computes a *t* value = 4.234 with 18 df and $p < 0.0005$, which means that the H_0 is not supported. Thus, it looks like Smartamine® may be the next super drug for learning and memory. But before the researcher starts thinking about submitting this work for publication in Science or Nature Neuroscience, perhaps other work needs to be done.

It is quite obvious in Figure 6.1 that there is considerable overlap in scores between the two groups with some of the animals receiving the experimental drug performing at the same level as subjects receiving the placebo. This is extremely common in animal experiments. Even though these results supported the alternative hypothesis, the previous study with a smaller sample size was not as supportive. Should those results be combined with these? Actually no, that would not be a wise decision, since one cannot be sure that all the conditions were the same for each of the studies. The prudent experiment would be to replicate the study again with an equal number of subjects per group (**balanced design**) or to use a repeated measure or crossover design. However, this study does demonstrate that with larger n's/group there is greater power and subsequently greater likelihood of observing a significant effect.

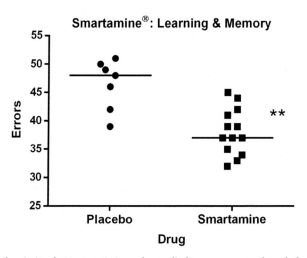

FIGURE 6.1 A simple *t*-test statistic can be applied to groups even though the n/group is not the same. **$p < 0.0005$ compared to the placebo group.

Paired t-Test: Testing the Means of Two Dependent Samples

Let us change the above experiment and make the two groups dependent samples. The researcher has decided to test the drug again but this time use each animal to test learning with both the new drug Smartamine® (S) and also the placebo (P). He will use only eight animals but will first give half the animals the placebo and subsequently retest the animals with the drug. The other half of the animals will receive the new drug first and then be retested with the placebo. In this way he can control for the effects of giving each drug first. Of course, the experimenter had to change the maze when the animals are tested a second time. The learning scores for each of the animals under the different conditions are given in Table 6.3.

Because the same animal is tested twice, it is a paired-comparison design (also called a **repeated measure design** or **crossover design**). One of the benefits of this type of design is that individual differences can be controlled. The learning for both the placebo condition and the Smartamine® condition is very similar to the experiment run above and the variance is very close for the two groups. However, the t value = 5.702 with 7 df and the $p < 0.001$. You might wonder why the large difference in p values. If you look at the data closely (Table 6.3) you will see that for every animal, the performance was better under the Smartamine® condition.

There are other types of paired comparisons, such as experiments where two animals are given the exact same treatment and animal No. 1 in group **A** is compared only to animal No. 1 in group **B**. In this situation one would again use a paired t-test to evaluate possible differences between group means.

TABLE 6.3   Paired t-Test: Smartamine®

Animal No.	Errors with placebo	Errors with Smartamine®
1 P–S	48	36
2 S–P	49	42
3 S–P	53	46
4 P–S	38	32
5 P–S	45	39
6 S–P	39	37
7 P–S	51	47
8 S–P	43	39
	45.9 ± 5.5	39.8 ± 5.1

COMPARING THREE OR MORE
INDEPENDENT GROUPS

If your experimental design has more than two groups and you want to compare the group means, the ANOVA is likely the most appropriate test. What this statistic does is examine the amount of variance in the dependent variable and try to determine from where that variance is coming. As with many statistical tests, there are some requirements that need to be followed:

1. The values to be evaluated must be quantitative (interval or ratio data).
2. The populations from which the samples are drawn must be normally distributed.
3. The samples must be independent of each other.
4. The groups must have variances that are relatively equivalent (homogeneity of variance).

The null hypothesis (H_o) for a simple one-way ANOVA is: $mean_1 = mean_2 = mean_3 = mean_k$.

(*Statisticians like to use the subscript "k" to indicate that there could be a lot more of something, like group means.*)

The alternative hypothesis (H_A) is that not all the means are equal and at least one is different from the others. This is very important, because **if your ANOVA rejects the H_o then all you can conclude is that a difference exists**. It does not indicate where that difference is or whether or not it is greater or smaller. That type of analysis requires the use of another statistic.

There are literally hundreds of different types of experimental designs and many of them require some rather unusual applications of the ANOVA statistic to properly analyze the results. A complete discussion of this topic is beyond the scope of this book and only some of the more basic designs will be discussed. For those interested in a more detailed discussion see Winer et al.[1]

COMPLETELY RANDOMIZED ONE-WAY ANOVA

This type of design is the simplest and involves the use of one independent or grouping variable (**factor**) and one dependent or quantitative variable. Statisticians use the term **factor** in an ANOVA because it can be assigned different levels. A level simply describes the different groups within the factor. For example, if you were interested in evaluating the effect of different antioxidants on cell firing in a hippocampal slice

TABLE 6.4   Hippocampal Cell Firing—Percent Change from Baseline

Antioxidant A	Antioxidant B	Antioxidant C	Antioxidant D
132	141	118	108
128	139	121	114
137	142	111	104
130	127	132	113
142	134	127	104
121	129	115	100
134	143	126	107
126	152	119	103
131.3 ± 6.6	138.4 ± 8.1	121.1 ± 6.9	106.6 ± 4.9

preparation, you might want to look at four different antioxidants. Each antioxidant would be a different group.

In this example presented in Table 6.4, the factor would be drug (antioxidants) and it would have **four levels**, corresponding to the four different antioxidants (A, B, C, D). The dependent variable would be the cell firing rate. The H_o is that the means of each group are equivalent and the H_A is that at least one of the means is different. It is a **completely random design** because the values for each antioxidant are independent from each other. It may be the case in this experiment that each of the antioxidants was tried in the same slice but for this example we will assume that it does not matter, only that there are eight different slice preparations used to obtain the data. The data are quantitative. The data in each group appear to be normally distributed. Is the variance in each group different? This can be evaluated with a test and is discussed below.

Because there are equal numbers of values for each group ($n = 8$/group), it is a **balanced design**. It is also important to point out that a one-way ANOVA **does not** require that the sample size be equal for each group. Thus you can have an unbalanced design and still be able to use this statistic. When these data are analyzed by a statistical program you will get a lot of different numbers depending on who the programmer was for that statistical package. A typical printout might be that shown in Table 6.5.

What do all of these numbers really mean? As a researcher you are primarily interested to know if your F value is significant or not. In the above experiment concerning antioxidants on the hippocampus, there was a significant effect of antioxidant on cell firing rate. The associated F score has a p value of <0.0001.

TABLE 6.5   ANOVA Table: Hippocampal Cell Firing

	df	Sum of squares	Mean square	F value	p value
Antioxidant	3	4551	1517	33.602	<0.0001
Residual	28	126	45		
Total	31	4677			

PARTITIONED VARIANCE

Statisticians usually say that the variation in the data is **partitioned** into two parts, i.e., a **within variance** and a **between variance**. The "within variance" looks at how much the values vary within groups and gives an indication of how a particular value in the group differs from the mean of that group. The "between variance" looks at how much values vary between the different groups and gives an idea of how different each group mean differs from the overall mean. These two partitions are then compared with one another. To do this, a software program calculates the "Sum of Squares" for both the within and between partitions, and then calculates what is called the "**Mean Square Error term**" (MSE). In Table 6.5 it is simply called the Mean Square. The MSE for the between partition ($MS_{between}$) is the larger number and it is calculated by dividing the Sum of Squares for all the values by the df (which is the total number of levels of the main factor—1). For example, in the hippocampal firing experiment above, there are four different levels of antioxidants. Thus the df is $4 - 1 = 3$. The MSE for the within partition (MS_{within}) (often called the residual) is calculated by dividing its Sum of Squares by the total df (total number of values—the total number of levels of the main factor; $32 - 4 = 28$).

To calculate the F score, the program uses this simple equation:

$$F = \frac{\text{Mean Square}_{between}}{\text{Mean Square}_{within}}$$

$$= \frac{\text{estimate of variance based on the group means}}{\text{estimate of the variance based on all the sample values}}$$

What exactly is this F score anyway? Since the ANOVA stands for analysis of variance, what it does is look at the variance between the different groups and decides whether or not they are the same or different. However, to do this it must also take into account how much variance there is within the groups. If you recall the unpaired t-test example above with the unbalanced design, there were some animals in the placebo group that performed as well as animals given the drug Smartamine® (Figure 6.1). That spread of different scores (variance) in the placebo group (and also in the drug group) is called the within-group variance for the placebo group (and the drug group). What the ANOVA does is calculate a value that is based on the **ratio** of the variance between the different groups under consideration divided by the variance that is observed within each of the groups. So you obtain the following:

$$F = \frac{\text{Between-group Variance}}{\text{Within-group Variance}}$$

The source of the within-group variance could simply be biological variance. In any given random sample it is quite possible to obtain values that are at the extremes of a distribution. You might take a single animal from your colony and test it in your maze and that animal does it perfectly the very first time. Is that animal representative of the colony? Well in some respect yes, but it does not mean that all the animals in the colony will perform the same way. If the F value is large it means that for that particular group sampling, the differences between groups is much greater than the differences within groups, and it is less likely that the differences observed are due to chance. In research, the standard rule is that the higher the F value the greater the probability that the differences are real.

Statisticians have generated a table (**Critical Values of F**) that is based on the probability distribution of different F values. What this table represents is the probability that a particular F value might occur by chance alone for an experiment with a specific number of experimental groups and a specific number of subjects. This is denoted by the df for both the between-group variance and the within-group variance. Most F distribution tables provide values for the probability of detecting an F score that is significant at the $p = 0.05, 0.01, 0.005$, and 0.001 levels. Although the computer programs automatically calculate the p value, you can check it by going to an F distribution table, which is easy to find on the web. The only problem with doing that is that these tables usually only provide the F values for significance up to the 0.001 level and it looks so much more impressive if you can say $p < 0.0005$ or $p < 0.0001$.

Statisticians often talk about a "Large or Big F." What a large or big F means is that the results **do not** support the null hypothesis but rather support the alternative hypothesis, that there is at least one difference in the group means. The larger F score equates to an increased level of significance supporting the alternative hypothesis. It is important to remember that the significance of the F score is dependent on the power, which is directly related to the number of subjects (values) in the analysis. If you look at an F table on the web, you will see that to obtain significance at even the 0.05 level, the required F value decreases dramatically as the df for the within MSE increases. This is directly related to the total number of values used and thus the number of subjects.

REPORTING ANOVA RESULTS

For a simple one-way ANOVA in a manuscript, it is required that you give the F value, the df, and the p value. Because an ANOVA has two different df associated with it, always report the between-groups df first. For the above experiment (Table 6.5): $[F(3,28) = 33.602, p < 0.0001]$. The use of brackets is simply to distinguish it from the parentheses around the df. Why is it important to show the df and the F value when reporting the results of an ANOVA? **Knowing the actual F value indicates the "robustness" of the analysis.** However, this can only be interpreted if you know exactly how many group means were involved in the analysis and how large the subject pool was. The numbers in the parentheses following the letter F in Table 6.5 show that the group size was 4 (number of groups: 1) and the total number of subjects used in the analysis was 32 (28 + number of groups). In the above experiment, the results could have been reported as $p < 0.05$ and it would be certainly correct. One would not know if it was actually significant at $p = 0.0499$ or $p = 0.0000499$. Does that really make a difference? An F score that is "barely" significant ($p < 0.05$) should be questioned because there is a greater chance that random sampling may have resulted in the difference. A significance of 0.0001 has a much greater probability that the results were **not** simply due to chance. In any experiment, when ANOVA results are reported, one should make a quick check to see if all the groups were included in the analysis and whether or not some of values were not included. In the above experiment, if the ANOVA results were reported as $[F(3,\mathbf{25}) = 33.602, p < 0.0001]$, you would immediately know that some of the values were dropped from the analysis because the df for the MS_{within} was low.

Thus far, the ANOVA indicated that the H_o should be rejected and the H_A is supported. This indicated that <u>at least one</u> of the antioxidant group means was different from the rest. There may actually be more than one but that requires additional statistical tests. Before talking about these tests it is important to discuss homogeneity of variance.

HOMOGENEITY OF VARIANCE

One of the basic tenets for using parametric statistics is **homogeneity of variance** and it is a key parameter. What does this actually mean? Variance is a measure of dispersion of data that will always exist in any type of biological experiment. In statistics, variance is the amount of dispersion around the mean of a given group of measurements. If one measured your reaction time on different days of the week, it would not always be the same because sometimes you might be more motivated than others or you might be distracted, tired, or even have consumed an interfering beverage. If we found the mean of your reaction time, we could also calculate the standard deviation of your different reaction times, and that would be a measure of the variance. In any given normal population of subjects, a random sample will have subjects that differ in regard to some biological variable. Using the reaction time example, some will have a mean time that is faster or slower than yours based on a variety of different variables.

Let us say that we are testing the effects of different herbal remedies on reaction time. We have randomly selected volunteers at the local pharmacy and asked them to participate (and they agreed!). We then recorded their reaction time over five trials, 30 min after taking one of four different herbal remedies. Naturally, all of the remedies were coded so we did not know who had what. Each group had 10 subjects and the reaction time data in milliseconds are shown in Table 6.6. The null hypothesis is that regardless of the herbal remedy, the different groups should show the same degree of variance regardless of the mean values. In other words, we are predicting that the variance in each group is unrelated to the size of any changes in group means.

A quick perusal of the table reveals that the numbers are not identical so there must have been some type of herbal remedy effect, but to test this, it is important to know if the variance in each group is the

TABLE 6.6   Herbal Remedies and Reaction Time

	CBRs	SupZ	HCE	KrO
	342.	257.	367.	271.
	365.	286.	368.	281.
	367.	223.	358.	267.
	334.	233.	373.	304.
	302.	246.	383.	225.
	325.	214.	376.	232.
	375.	237.	363.	256.
	390.	205.	372.	301.
	372.	289.	379.	247.
	334.	254.	380.	322.
Mean	351	244	372	271
SD	27.3	28.1	7.9	31.9

same. It is too bad that we did not have a before and after test, or did not test the same group of subjects with each of the different remedies. There are a number of different tests that can be used to determine if each of the groups, regardless of their means, has equivalent variance. Most statistical software programs can run one or several of the tests listed below.

Bartlett's Test

This test published in 1937 by M.S. Bartlett[2] evaluates whether or not a group of values comes from a normal distribution. It does not really test for variance but it is often used as a test of homogeneity of variance. If this test has a significant p value, it is an indication that one of the groups has a significantly different variance. This is a very common test used by commercially available statistical packages along with the Brown–Forsythe test.

Brown–Forsythe Test

This statistic named after Morton Brown and Alan Forsythe[3] uses the median instead of the mean in its analysis and consequently provides greater sensitivity to departures. If the p value is significant, it is a very good indication that there is heterogeneity and not homogeneity of variance.

Levene's Test

This statistic created by Howard Levene[4] actually tests the hypothesis that the mean variances in each group are equal using an alpha (α) of 0.05. It is not as sensitive to departures as the Brown–Forsythe test. If the p value is significant, it indicates a significant problem in the variance between groups.

O'Brien's Test

This statistic, created by Ralph O'Brien,[5] is very similar to the Brown–Forsythe test and provides good sensitivity. It is often found in commercially available statistical packages.

Hartley Test

Herman Otto Hartley[6, 7] devised this relatively simple statistic test. This test uses what is called the F_{max} statistic and assumes independent random sampling. It is unclear why it is not included more in commercial statistical packages.

On the above data in Table 6.6 here are the results of two tests of homogeneity of variance:

Bartlett's test

$$p < 0.005 = \text{significant difference in variance}$$

Brown–Forsythe test

$$p < 0.05 = \text{significant difference in variance}$$

So what variance is different? A simple way to tell is to graph the data using box and whiskers (Chapter 4) and observe if the boxes are the same or if there is one that looks out of place. In Figure 6.2, it appears that the HCE group has very little variance and is not even close to the other three groups.

So what does this mean? There are two possibilities that could account for the differences observed. It may be the fact that the HCE group is simply not representative of the distribution, and somehow during the recruitment there was a significant bias that may have interfered with the testing. Alternatively, the HCE herbal remedy has a profound effect on this random sample of the population and significantly slows down the reaction time. Even if this were the case, one would expect that the variance should be more equivalent to that of the other groups. In this particular case, careful review of the experimental notes revealed that all of the individuals in the HCE group were ≥69 years old, and age may have been a confounding factor in the study. The fact that the group

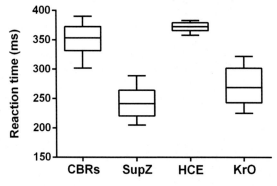

FIGURE 6.2   This box and whisker plot shows the data for the reaction time from individuals that were treated with different herbal compounds. Three of the four groups appear to have similar variance.

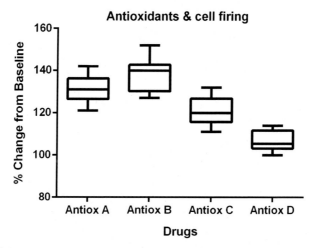

FIGURE 6.3 Although the group medians appear to be different there does not appear to be a large difference in the variance between groups as shown in this box and whisker plot.

variances did not meet the requirement for homogeneity would preclude the use of a parametric one-way ANOVA. One could still analyze with a nonparametric ANOVA (Chapter 8).

Reviewing the variance between groups prior to carrying out a parametric statistic could prevent a misleading interpretation of the data. As a cautionary note, **outliers** (Chapter 9) within a set of data can significantly influence the results of all of the above tests. It is thus wise to graph the data if the homogeneity of variance tests shows significant results (Figure 6.3). In our example of antioxidants and hippocampal cell firing (Table 6.4) was there homogeneity of variance?

If the data were graphed using box plots, and there does not appear to be any significant outliers, and the variance between groups looks pretty even, then there is probably homogeneity of variance. Subjecting the data to both the Brown–Forsythe test ($p > 0.7$) and the Bartlett's test ($p > 0.6$) confirmed this for the data in Table 6.4 and thus this requirement for the use of the ANOVA was not violated (Table 6.7).

So we are now trying to answer the question as to which of the mean(s) in the ANOVA is (are) significantly different. The one-way ANOVA resulted in $F(3,28) = 33.602$, $p < 0.0001$, which is a very big F score. **This next analysis is perhaps the most controversial**.

TABLE 6.7 Hippocampal Cell Firing—Percent Change from Baseline

Antioxidant A	Antioxidant B	Antioxidant C	Antioxidant D
131.3 ± 6.6	138.4 ± 8.1	121.1 ± 6.9	106.6 ± 4.9

MULTIPLE COMPARISONS

Planned Comparisons (A Priori)

When someone designs an experiment, he or she usually has an idea of what hypothesis he or she wants to test or what different hypotheses he or she believes will be interesting. The preferable way to carry out an experiment is to state, prior to collecting any data, what specific comparisons are to be made, rather than determining this after the data have been "reviewed." **Planned comparisons**, also known as a priori comparisons, have more power than **post hoc** comparisons because they do not correct for the probability of making a Type I error. However, the number of a priori comparisons should not be extensive and should be logical extensions of the experimental rationale. Planned comparisons can use a number of different statistics including the t-test and the Tukey's Honestly Significantly Different test (**HSD**).

Post Hoc Comparisons (A Posteriori)

Most experimental situations look at "after the fact" or post hoc comparisons. First, was the F score significant? If the answer to this is <u>no</u>, then you probably do not need to investigate any further. However, there are some cases where it is ok to investigate further, but this is a topic for a lot of beer and pizza. Interested individuals can chat with their friendly statistician or look at the discussion in Kirk.[8] If the F score is not significant it means that there were not enough differences between the means, given the amount of variance, to warrant rejection of the null hypothesis. That is, changes in the levels of the independent variable did not have any differential effects on the dependent variable in which you were interested. The variance between groups was simply too great. For example, you were sure that the drug you were administering would make the rats perform "better" when learning the maze, and you even tried four different doses. The overall ANOVA was not significant and thus the drug probably does not have any biologically significant effect.

So you ran the ANOVA and the printout "said" the generated F score was significant, thus rejecting the null hypothesis. This indication of significance simply means that <u>at least two</u> of the group means were probably significantly different from each other. To find out which means changed, you will have to carry out some "post hoc" testing or do what is called "**data snooping**." The major question here is that you are probably not sure which test is "the best," or maybe the software you are using gives several different choices. Now you are trying to figure out which test to choose.

Depending on the author of a particular statistics text, one can make a case for just about each and every one of the various (30 or so) multiple comparison procedures (**MCPs**). It seems that everyone has some "better" way to data snoop. Here is a brief description of some of the most common options and why some are better than others in a given situation. You have to think about each of the different pairwise comparisons as a single hypothesis. That hypothesis would say that there is no difference between the two groups. If you have a number of groups, then you have a number of different hypotheses. In an experiment with four independent groups, there can be a total of six different hypotheses tested: A vs B; A vs C; A vs D; B vs C; B vs D; C vs D.

Here is an example of a rather simple experiment with four different groups (Table 6.8). You have a null hypothesis that the drug Smartamine® does not make the subjects perform any different than a cohort given saline alone. But if it does, then you probably would like to know which of the doses was responsible for the difference. You might also want to know if the change is dose dependent. In other words, does giving a subject 25 mg/kg alter the behavior differently than a 10 mg/kg therapy?

Look at the F score in Table 6.9. This value indicates that the overall ANOVA is very robust and a biostatistician would say you have a big F.

Now look at the group means in Table 6.10. There appears to be a big difference between those injected with the drug and the saline-treated group. So are they significantly different?

The simplest way to do this is to use a t-test. After all, a t-test is the same as a one-way ANOVA with only two groups. If you decide to do this,

TABLE 6.8 Smartamine® Dose-Dependent Changes

Saline	Smartamine® 5 mg/kg	Smartamine® 10 mg/kg	Smartamine® 25 mg/kg
40	44	24	36
63	34	20	41
58	21	29	53
61	29	30	49
51	36	39	37
43	50	18	32
48	32	22	46
43	33	37	50
44	25	19	48
60	42	38	45

TABLE 6.9   ANOVA Results: Smartamine®

F score	df	p value
15.80	3,36	0.0001

TABLE 6.10   Smartamine®
Dose-Dependent
Changes

Group	Mean
Saline	51.1
5 mg/kg	34.6
10 mg/kg	27.6
25 mg/kg	43.7

one must use what is called a **correction factor.** The reason for this is really quite simple. The more tests you run comparing groups, the greater the likelihood that you will obtain significance by pure chance. Probability says that you will eventually make a Type I error. You will conclude that something is significant when in fact it is not.

Here is an interesting table showing the probability of rejecting the null hypothesis with an α set at 0.05 and increasing the number of groups (Table 6.11). With only two groups there is a single comparison, but with three groups there are three comparisons (1 vs 2; 1 vs 3; 2 vs 3). As you increase the number of groups, the number of comparisons also increases.

This means that if you have an experiment with six different groups, and you compare each group with all the other groups, there will be 15

TABLE 6.11   Probability of Rejecting Null Hypothesis

Number of groups	Number of comparisons	Probability of incorrectly rejecting the H_o at 0.05 with a *t*-test
2	1	5%
3	3	14%
4	6	26%
5	10	40%
6	15	50%
7	21	64%

different comparisons. With 15 different comparisons, there is a 50:50 chance that you will reject the null hypothesis at least once, and say there is a significant effect when in fact there is not. In other words, there is a 50% probability that you will make at least one Type I (false positive) error. Naturally, you will not know which of those comparisons you think are significant really are not.

Your friendly biostatisticians, if you talk with them, will mention the term **"familywise error rate (FWER)."** Here is what that means. A group of hypothesis comparisons that are made in one experiment is called a **"family"** of comparisons. In the example above with Smartamine®, one of these comparisons in the family could be: "I hypothesize that animals given 25 mg/kg of Smartamine® will perform equivalent to animals given 5 mg/kg Smartamine®." If the experiment is very complex, the family of comparisons could be the group of comparisons in only one major part of that experiment. What I always find fascinating is that your friendly biostatistician cannot really give you any hard and fast rule about which comparisons should or should not be included in a family. A good rule to follow is that any comparison made between groups that were included in your ANOVA is part of the hypothesis family. As you make more and more hypothesis comparisons, the size of the family increases. As the size of the family of comparisons increases, the chance of making a Type I error also increases, and this is the familywise error rate. The FWER simply means that the probability of making a Type I error increases as you make more comparisons. It is the probability that within a family of hypothesis comparisons there will be at least one comparison that is determined to be significant when it should not be, or falsely rejecting the null hypothesis. You are said to control the FWER if it can be predicted that **not more than one** of your comparisons falsely rejects the null hypothesis. To *reduce* the risk (but not totally eliminate it) there are protection procedures or correction factors. There are several different MCPs that have these protection procedures "built in." Incidentally, your friendly biostatistician may call comparisons **"contrasts."**

What is the issue that needs to be considered when doing this type of data snooping? You have collected your data and now you want to know if something is significant or not. In other words, you may have evidence to reject the null hypothesis but you do not want to reject it if the evidence is not really there—this would be a Type I error. By convention, the FWER is set at $p \leq 0.05$. If you only run a single test, such as an experimental group versus a control group, then an increased risk of a Type I error can really only occur if you have somehow breached one of the theoretical rules that are associated with that particular test, such as homogeneity of variance. However, if there are multiple different hypotheses in the family to be tested, then the risk of making that Type 1 error is increased beyond the initial $p \leq 0.05$ unless some type of precautionary measure is used. This is the basis of the MCPs.

MULTIPLE t-TESTS

Let us look at the multiple t-test procedure. The multiple t-test procedure **does not** have any protection built in. If you want to use this statistic as a follow-up to the ANOVA, then it is necessary to insert a correction factor. Let us say that you have set your α level at 0.05 for determining significance between groups. What would happen if you simply did not use a correction factor for multiple comparisons and just used multiple t-tests? You would obtain the results shown in Table 6.12.

You would conclude that in fact Smartamine® significantly improved performance at all the doses used, and both the 5 and the 10 mg/kg doses, while not different from each other, were significantly different compared to the 25 mg/kg dose. In all, there were six comparisons in this family, and five of the six rejected the null hypothesis. But Table 6.11 shows that with six comparisons there is a 26% chance of making at least one Type 1 error, so perhaps that is the case with this analysis.

The simplest and most common correction factor used with the t-test is the **Bonferroni procedure** or **Dunn's multiple comparison procedure** published by Olive J. Dunn.[9] This is often referred to as the Bonferroni or the Dunn procedure. It got the name Bonferroni procedure because the published tables are based on the Bonferroni inequality. One simply takes the α level (in this case 0.05) and divides it by the number of comparisons that are made (in this case 6), which results in a new α value of 0.0083. That means that to be significant at the original α level of 0.05, the t-test significance must be equal to or less than 0.0083. A corrected α of 0.009 would **not be significant** at the 0.05 level. In the above set of comparisons, only the first, second, and sixth comparisons would be significant. It is

TABLE 6.12 Multiple t-Tests without a Correction Factor

No.	Comparisons	t value	df	p value	Significant?
1	Saline, 5 mg/kg	4.212	18	0.0005	Yes
2	Saline, 10 mg/kg	6.223	18	0.0001	Yes
3	Saline, 25 mg/kg	2.108	18	0.0493	Yes
4	5, 10 mg/kg	1.840	18	0.0823	No
5	5, 25 mg/kg	2.570	18	0.0193	Yes
6	10, 25 mg/kg	4.758	18	0.0002	Yes

TABLE 6.13 Multiple *t*-Test with the Bonferroni Correction Factor $\alpha = 0.05$

No.	Comparisons	*t* value	df	*p* value	Significant?
1	Saline, 5 mg/kg	4.212	18	0.0005	Yes
2	Saline, 10 mg/kg	6.223	18	0.0001	Yes
3	Saline, 25 mg/kg	2.108	18	0.0493	No
4	5, 10 mg/kg	1.840	18	0.0823	No
5	5, 25 mg/kg	2.570	18	0.0193	No
6	10, 25 mg/kg	4.758	18	0.0002	Yes

probably quite obvious that as the number of comparisons increases, it is incrementally more difficult to obtain significance. With six comparisons, the *t* value would have to be 2.795 or higher to be significant at the 0.05 level (see Dunn[9]) (Table 6.13).

A statistician, Zbyněk Šidák, developed a modification of the Bonferroni/Dunn correction procedure (**Šidák correction** or **Dunn−⊕ Šidák correction**) that is sometimes used.[10] This correction method is a little more liberal than the Bonferroni/Dunn but still extremely conservative and should be used with caution for greater than three comparisons (Table 6.14).

A conservative test will detect fewer significant differences and make more Type II errors. A liberal statistic will detect multiple differences and make more Type I errors. Even with three comparisons, the new α value would be $0.05/3 = 0.017$. It is extremely important to remember that as one decreases the chances of making a Type I error, they increase the chance of making a Type II error and ultimately reduce the power of the statistic.

TABLE 6.14 Comparison of Bonferroni and Šidák Correction Factors with Multiple *t*-Tests

No.	Comparisons	*t* value	df	*p* value	Bonferroni	Šidák
1	Saline, 5 mg/kg	4.212	18	0.0005	0.05	0.01
2	Saline, 10 mg/kg	6.223	18	0.0001	0.05	0.01
3	Saline, 25 mg/kg	2.108	18	0.0493	NS	NS
4	5, 10 mg/kg	1.840	18	0.0823	NS	NS
5	5, 25 mg/kg	2.570	18	0.0193	NS	NS
6	10, 25 mg/kg	4.758	18	0.0002	0.05	0.01

There are two other modifications to the Bonferroni/Dunn method that have been called the **Holm—Bonferroni**[11] (sometimes called the **Holms test**, or **Bonferroni—Holm test**) and the **Hochberg** procedure,[12] which are better than both the Bonferroni/Dunn and the Šidák methods. What both of these statistics do is rank the p values generated from t-tests in ascending or descending order, and the p values are then evaluated sequentially.

With the Holm test, the original α (0.05) is divided by the total number of comparisons (0.05/6) and the value (0.008) compared to the t-test p value (0.0001). If the original value is lower than the correction value it is considered significant. The next p value is evaluated; only this time the original α is divided by the number of remaining comparisons (total comparisons minus 1). The new comparison α is $0.05/5 = 0.01$. Once a value is reached that is not significant, the remaining comparisons are considered not significant. The major change with the Holm—Bonferroni is that it takes into account how many comparisons have already been analyzed and becomes less conservative with each additional comparison. There is also a Holm—Bonferroni Šidák test (often called **Holm—Šidák t-test** or **Šidák—Holms test**), which is a little less conservative and more powerful than the Holms test. It is often recommended for making multiple comparisons and very similar to the Holm—Bonferroni method. The Šidák—Holms test is slightly less conservative than the Holm—Bonferroni method and does a good job of controlling the FWE. Many commercially available statistics packages actually use the Holm—Šidák test and call it the Holms test for multiple

TABLE 6.15   Comparison of Holm—Bonferroni and Hochberg Factors

Holm—Bonferroni			Hochberg		
0.001	0.05/6 = 0.008	Significant	0.0823	0.05/1 = 0.01	Not significant
0.007	0.05/5 = 0.01	Significant	0.0215	0.05/2 = 0.025	Significant
0.0128	0.05/4 = 0.0125	Not significant	0.0135		Significant
0.0135		Not significant	0.0128		Significant
0.0215		Not significant	0.007		Significant
0.0823		Not significant	0.001		Significant

comparisons. It has the added feature that it can be used when the homogeneity of variance and the population normality are questionable.

With the **Hochberg procedure**, the opposite is carried out beginning with the least significant. Once a comparison is determined to be significant, all remaining comparisons are also deemed significant. The value here is the fact that unlike the Bonferroni/Dunn procedure, the comparisons are adjusted depending on the remaining possible comparisons. So, do these two methods always give the same results? Actually they do not and the Hochberg is a little less conservative. Table 6.15 shows a comparison of t-test outcomes on a set of data with both the Holm–Bonferroni and the Hochberg procedures.

With the Hochberg procedure five of the comparisons are significant and with the Holm–Bonferroni only two. The Hochberg is found in several commercial statistical packages. This procedure is more powerful than the Holm–Bonferroni but one cannot obtain confidence intervals.

FALSE DISCOVERY RATE

Although the above procedures have been used to control for the FWER, there are certainly instances in which an extremely large number of comparisons can be made. For example, in cases where neuroimaging is used, multiple different very small regions of the brain are compared. If one were to use the Bonferroni correction method, the adjusted α level would be so small that it would be virtually impossible to detect any significant difference. A relatively new approach to this problem is to try and control the *proportion* of false positives that will be detected.

In this procedure, which is less strict than those that control the FWER, some false positive findings are allowed. Procedures that control the FWER attempt to prevent even one false positive to occur, while the false discovery rate (FDR) controls the proportion of false positives to some reasonable fraction of all the tests. It subsequently allows for the possible discovery of important effects that could indicate future, more controlled testing.

Here is how it works. First, one assumes that there will be a number of false positives and thus the p values associated with these results will have a different risk associated with them. Suppose you have 1000 different comparisons with some having a significant p value. A certain percentage of these are false positives. The researcher then sets a limit on the proportion that will be accepted as a significant discovery and this value is called **Q**. If one sets the Q value to be 5%, then 50 of these will be false discoveries. The actual procedure to do this is beyond the scope of this text but many statistical packages feature FDR as an option. Its major advantage is that it can be used to test situations where there are a large number of comparisons (contrasts).

COMMON POST HOC TESTS

There are many other different *post hoc* MCPs that have protective procedures and the choice is often very complicated. The selection of which procedure to use is probably the **most debated**. I should also say that the **actual choice** of which post hoc test is used is probably also the **most abused** in statistics. In fact, the field of MCP is so broad that to do it justice is beyond the scope of the present discussion. There are simply so many different published papers on this topic that a complete review would probably never get accepted because of the different opinions of the different statisticians. If you have some extra time and really want to be both amused but totally confused, gather three statisticians together and bring up the topic of MCPs. Just remember that I warned you. That being said, the following section will present some guidelines (my bias) for one's choice for some of the most common "data snooping" procedures. Many of these post hoc tests can be run simply by using the ANOVA analysis table that shows the MSE terms and the df and knowing the number of subjects per group.

Fisher's Least Significant Difference Test

One of the earliest post hoc MCPs was the least significant difference test (LSD) described by R.A. Fisher in 1935 and then later published in 1949.[13] This test is also known as Fisher's LSD and Fisher's Protected LSD. Actually these names are for the same test and the word "protected" simply means that one must have a significant **omnibus F test** (ANOVA F score) before using it. If you do not obtain a significant F score with the ANOVA, then you have to conclude that the group means are equal so no need to investigate further. The Fisher's LSD (or **PLSD**) differs from the multiple t-tests with the Bonferroni correction factor in that it uses the pooled variance from all of the groups used in the ANOVA, which gives a more accurate value of the standard deviation. A t-test only uses the variance of the two groups that are being compared. The PLSD statistic really does not account for multiple comparisons but is very powerful. If there are only three groups in the experimental design it protects against a Type I error just as well as any of the other MCPs but has more power. The test begins to break down in terms of protection when there are four or more groups that are being compared. In an attempt to correct for this, A.J. Hayter[14] proposed a modified LSD test (**Fisher–Hayter procedure**) that uses the studentized range distribution instead of the Student's t-distribution in calculating the LSD value. In addition, it treats the design as one less group when looking up the value in the studentized range distribution table. With this modification, the Fisher–Hayter is more conservative than the Fisher's LSD but still maintains good power. It has more power than many of the other MCP including the Tukey's HSD. Both the PLSD and the Fisher–Hayter can be used with both equal and unequal numbers of subjects per group, but this test requires a significant F test first. Like the PLSD, one cannot construct confidence intervals with the Fisher–Hayter.

The Fisher's LSD test is very good as long as the overall F score is significant but is considered to be extremely liberal and many statisticians believe it should not be used in complex experiments involving four or more groups. That is probably a good recommendation but then if you have a simple ANOVA with a very robust F score it is a good test. Statisticians are correct in saying that overall, it does not control the FWER directly, but it does so indirectly by demanding that the F score be significant. The Fisher–Hayter makes up for some of the problems some statisticians have with the PLSD, and has a nice mix of protection and power when all possible comparisons are being made. It is less powerful when only a few of the possible comparisons are carried out such as with a planned set of comparisons. In this case, the Bonferroni/Dunn or the Dunn–Šidák procedure is better. Both the PLSD and Fisher–Hayter require homogeneity of variance and certainly can be used following an ANOVA with a robust F score. There is a good discussion of the Fisher–Hayter and the Fisher's LSD in Kirk.[8]

Tukey's Honestly Significantly Different Test

The **HSD** was constructed by John W. Tukey[15] in response to the PLSD procedure, with the intent of keeping the risk of a Type I error low. It is what can be called an **alpha-adjusted test**. What this means is that the probability of committing a Type I error is reduced by making the H_o rejection region smaller, which reduces the FWER. Consequently, it is more conservative than the PLSD and Fisher–Hayter tests. The HSD statistic is also known by other names: Tukey's Range test, Tukey's method, Tukey's HSD, and Wholly Significant Difference (**WSD**) test. It is not the same as the Tukey–Kramer test.

This MCP statistic requires that the **number of subjects per group be equal** and there is homogeneity of variance. If there are only a few planned comparisons, the Tukey's HSD procedure is excellent and preferred over the Dunn procedure because it is more powerful. The **Tukey–Kramer method (TK)**, developed by C.Y. Kramer,[16] is very similar to the HSD but includes an adjustment for unequal subjects in the different groups. Consequently, this does have an effect on the outcome. Both the Tukey's HSD and the Tukey–Kramer tests are based on the studentized range distribution, similar to the Fisher–Hayter LSD test. Unlike the Fisher's PLSD test that provides an actual p value, the HSD sets an experimentwise error rate by establishing a critical value for "Q" at a specific α (e.g., 0.05) for the entire family of comparisons. This is an important point. If the mean difference exceeds the critical value, it is considered significant only at the level defined by the investigator. For example, if you have an experiment with the following group means (51.5, 39.4, 33.8, 29.3) and the critical difference between means at $\alpha = 0.05$ level is 9.993, then some of the comparisons will be significantly different way beyond the 0.05 level but you will not know this. The Tukey's HSD is not a protected test and can be run even without first doing an ANOVA. However, in this scenario, the number of comparisons should be limited.

The Tukey–Kramer test has the same basic assumptions as the Tukey's HSD but uses the **harmonic mean** (Chapter 3) to control for the unequal sample sizes. It is a bit more liberal than the HSD even though it also relies on the studentized range distribution. Nevertheless, both tests are conservative and strongly protect against a Type I error. Both of these tests are also sensitive to differences in group variance. A confidence interval can be calculated for both the HSD and the TK, which differs from the Fisher–Hayter procedure.

Student Newman–Keuls and Duncan Multiple Range Tests

These two tests are almost identical. The **Student Newman–Keuls (SNK)** was first proposed by Dennis Newman[17] and then popularized by Matthijs Keuls,[18] a Dutch horticulturalist. David B. Duncan[19] developed a modification of the SNK, called the **Duncan Multiple Range test (DMR)** that was to have greater power. Like the Tukey's HSD, the SNK calculates a "Q" value that is used for comparisons. These differ from the Tukey in that a different critical Q value is obtained for each comparison so it does not maintain the same familywise error rate for all comparisons. Both the SNK and the DMR tests use a stepwise MCP in which the means are listed in rank order and then systematically compared. For the SNK there is a prescribed sequence for comparisons that needs to be performed. First the highest and lowest means are compared and then the highest and second lowest, then third, and so forth. The critical difference for obtaining significance changes continuously and is based on the studentized range distribution. The major problem is that if there are many means to be compared the probability of finding a false positive (Type I error) increases dramatically. In other words, it does not control for the FWER. The SNK, like the Tukey's HSD, requires that the sample sizes be equal. The DMR is almost identical to the SNK, but instead of using the studentized range distribution, a modified distribution is used (Duncan's New Multiple Range distribution), making the test more liberal. While these two methods were once popular, both the SNK and the DMR are now considered as **tests to avoid** because they do not protect against the Type 1 error, especially with many pairwise comparisons. However, if there is only a very small subset of the possible contrasts and **they are planned** (a priori), the SNK is an adequate test. If showing a confidence level is important, then SNK cannot be used.

Games—Howell Procedure

The Games—Howell procedure **(GH)**, published by P.A. Games and J.F. Howell,[20] was specifically designed for instances when both the variance and the samples size were unequal. However, it can also be used for cases where the variance and sample size are equal, but loses some of its conservative properties. It uses the studentized range distribution and is a protected test dictating that it should only be used following a significant omnibus F test. Closely associated with the GH procedure is the **Tamhane procedure**[21] that uses the Student's t-distribution based on Šidák multiplicative inequality. In several different comparison studies, the GH procedure was deemed the most appropriate under conditions of unequal variance and unequal sample size. It is a test that is considered more conservative than the Tukey—Kramer but it can be used with very small sample sizes (e.g., <6). Overall it is a good post hoc test especially when the population variances are known to be unequal.

Scheffé Procedure

The Scheffé procedure is a very popular MCP that was developed by Henry Scheffé at Columbia University[22] and is used to make any possible contrasts among a family of comparisons. It is an extremely conservative test that does not require equal sample size. Often it is used after one finds a significant ANOVA and some authors believe that it should **only** be used after an omnibus F. Although it can handle some heterogeneity of variance, it works best if the variances are equal. It holds the FWER at a constant, regardless of the number of comparisons being made. Because it controls for an infinite number of contrasts, it is less powerful than the Tukey's HSD. It is important to remember that this is **not** a test for a priori hypothesis comparison because it was designed to be exclusively a conservative post hoc test. It also should not be used to make all pairwise comparisons, especially if there are more than six because it will dramatically lose power. However, this procedure is really recommended to be used to make some very elaborate comparisons in elaborate designs such as comparing groups 2 and 4 versus the means of groups 1, 3, and 5. Because of its extreme conservative properties, it is not recommended for many animal experiments where the investigator wants to make all pairwise comparisons. It is the least statistically powerful MCP and thus does not protect against a Type II error.

A final word of caution about using the Scheffé after an ANOVA. It is quite possible to obtain a significant F with the ANOVA, indicating that at least one of the group means is different, and then none of the contrasts are significant by the Scheffé statistic. Conversely, if the ANOVA is not significant then it is pointless to run a series of Scheffé tests.

Brown—Forsythe Procedure

This test developed by Morton Brown and Alan Forsythe[3] is a modification of the Scheffé procedure and like the Scheffé procedure can be used with unequal sample sizes and there is some question about the homogeneity of variance. It is best used when only a very few a priori comparisons have been planned. Most statisticians believe that the Games—Howell is a much better choice for use of heterogeneity of variance and unequal sample size.

Dunnett's Test

The Dunnett's test has a very specific use, the comparing of all treatment groups to a single control group such as a vehicle or placebo group. It was devised by Charles Dunnett[23] in 1955 for the express purpose of making multiple comparisons to a single group. Since the total number of contrasts is limited by the experimental design, and not all possible comparisons are made, it has greater power than the other tests noted above. It is actually a modification of the *t*-test and has some relationship to the Fisher's LSD procedure. The original Dunnett's test required that the variance of the control group be equal to the variance of the difference treatment groups. In 1964, Dunnett published a modification so that it could be used with unequal variance.[24, 25] There is a caveat to the use of this test that may give some a few nightmares. The test was designed to compare totally independent groups with a single control group. If one is running a drug study, such as the Smartamine® study (Table 6.8) applying the Dunnett's test is not correct. The reason is that when you are using different doses of the same drug, <u>each of the different doses is not considered as totally independent groups</u>, because the treatment groups are really related to the same drug. However, if three totally different drugs were being tested, then the Dunnett's test would be appropriate. It is also important to understand that this statistic is only used with a single factor experiment and **cannot be used after a multifactorial ANOVA**.

HOW TO CHOOSE WHICH MCP (POST HOC) TO EMPLOY AFTER AN ANOVA

It is extremely difficult to tell someone what specific MCP they should use because there are many factors to take into consideration. If the comparisons are planned (a priori), then a *t*-test with the Holm–Bonferroni correction factor or the Šidák correction is quite appropriate because it is not too conservative and still has power. On the other hand, if the family of comparisons is indeed data snooping (post hoc) tests then there are a number of factors to consider.

- How many comparisons/contrasts do you intend to make.
- Are the sample sizes equal among all groups.
- Is the sample size large or small (≤ 4).
- Is the variance roughly the same for each group.
- Are you more interested in protecting against a Type I or a Type II error.
- How conservative/liberal do you want to be.
- Do you need confidence intervals.

You cannot run multiple different MCP on your data, examine the outcomes, and subsequently decide which one gives the results that best fit your idea of how the experiment should have turned out. This is a form of ***p*-trawling**. This means continuously evaluating the data with a different statistic until a $p < 0.05$ is obtained for the comparisons that support one's overall hypothesis. An alternative to this is ***p*-hacking**, which is running an experiment, analyzing the data, and then determining that the sample sizes need to be larger to obtain significance at the $\alpha = 0.05$ level of significance, and then adding more subjects.

A commercially available statistical program cannot make the decisions for you. Statistical software programs will run whatever MCP you ask it, whether it is appropriate or not.

Possible recommendations following a significant *F* test:
If you have

- equal number of subjects/group
- the sample size is >6
- homogeneity of variance
  - Tukey–Kramer
  - Fisher–Hayter (no CI possible)

- equal number of subjects/group
- the sample size is >6
- **heterogeneity** of variance
  - Games–Howell

- **unequal** number of subjects/group
- sample size is >6
- homogeneity of variance
  - Tukey–Kramer
  - Fisher–Hayter (no CI possible)

- **unequal** number of subjects/group
- sample size is >6
- **heterogeneity** of variance
  - Games–Howell

- equal or unequal number of subjects/group
- **sample size is ≤ 6,**
- homogeneity of variance or heterogeneity of variance
  - Games–Howell

- equal or unequal number of subjects/group
- homogeneity of variance
- **only making comparisons to a single control group**
  - Dunnett's

- equal or unequal number of subjects/group
- homogeneity of variance
- comparing **only 3 groups**
- **want to protect against at Type II error**
  - Fisher's LSD (no CI possible)

Most of the time one really wants to protect against a Type II error and if the variance looks good and the n/group is the same, then run a Fisher–Hayter. If that is not available on your statistical package (and you do not want to compute it by calculator) then run a Games–Howell. If you are simply comparing independent means to a control group and not interested in individual cross-mean comparisons then it is the Dunnett's test.

ONE-WAY REPEATED MEASURES (WITHIN-SUBJECT) ANALYSIS OF VARIANCE

A one-way repeated measures ANOVA is very similar to a "regular" one-way ANOVA with the exception that the measures across the independent variable are derived from the same subjects and not from independent subjects. Consequently, the data must be analyzed differently because of the relationship of the different scores to one another. It can be thought of as an advanced dependent *t*-test. A repeated measures ANOVA is sometimes also called a **within-subjects ANOVA** or an **ANOVA for correlated samples**. As with a simple one-way ANOVA, the repeated measures variety still has one independent variable (with multiple levels) and one dependent variable. A major advantage of using a repeated measures analysis is that it often reduces the amount of biological variance, which subsequently increases power. Often in these types of experimental designs, a particular subject serves as its own control. This can often significantly reduce the total number of subjects that need to be evaluated. A relatively common use of a repeated measure design is to evaluate a specific response over time, such as a change in a serum biomarker following exposure to a specific stimulus. It is very important that each subject has values for each cell of the matrix. If a single value is missing, most statistical programs will eliminate that subject from the total analysis. While this type of ANOVA normally requires a balanced design, there are some statistical tactics that can be used to get around this. A complete discussion of these tactics is beyond the scope of the present discussion (Chapter 9) but treated at length by Winer et al.[1] Alternatively you can consult with your friendly neighborhood biostatistician.

Here is an example of this type of design and analysis (Tables 6.16 and 6.17). A researcher is interested in determining if a specific novel compound has a time-dependent effect on cell firing rate in the ventrobasal complex of the thalamus. Six young naïve male rats are to be evaluated at four different times post injection. A recording electrode is implanted in the ventrobasal complex and thus firing rate can be studied over time.

In this particular experiment, the F value is very large and consequently the p value is very small, indicating that there was at least one time point in which the novel compound had a significant effect on the cell firing rate. The next question is to determine which time point or

TABLE 6.16   Cell Firing Rate in Ventrobasal Thalamus

Subject No.	Baseline	30 min	45 min	60 min	90 min
1	26	40	49	57	62
2	24	21	38	63	47
3	19	37	50	79	56
4	21	27	44	48	44
5	27	38	32	67	58
6	22	42	49	72	57

TABLE 6.17   ANOVA Table: Cell Firing Rate

	df	Sum of squares	Mean square	F value	p value
Subject	5	629	126		
Cell firing	4	6265	1566	35.971	<0.0001
Cell firing * subject	20	871	43.5		

points contributed to this big F value. This now comes back to the basic question that the experiment was designed to answer. If the question is whether or not there is a change in baseline activity at various times after introduction of the novel compound, the appropriate post hoc test would be the Dunnett's MCP. This particular test compares the cell firing rate at each time point with only the baseline and ignores comparisons between the different time points. For these data, this post hoc analysis revealed a significant increase at 45, 60, and 90 min post injection but not at 30 min.

Here is another example of a one-way repeated measure ANOVA (Table 6.18). An investigator evaluated how swim speed might differ in a group of six different young adult male rats on different days as a result of a change in water temperature. Water temperature might be a significant factor in the learning of a task such as the Morris Water Maze. The H_0 is that the swim speed will be the same regardless of the water temperature.

The overall analysis (Table 6.19) states that water temperature had a significant effect on swim speed and this effect was quite robust. Post hoc analysis must be applied to determine which of the group means are significantly different.

Unlike the first experiment involving the cell firing rate in the thalamus, this study did not rely on a baseline for comparison. Instead, four

TABLE 6.18   Swim Speed: Effect of Water Temperature (°C)

Animal	20°	24°	27°	32°
1	39	38	34	33
2	29	25	20	20
3	36	37	29	24
4	25	29	18	19
5	31	27	24	22
6	34	33	30	31
Mean ± SD	32.3 ± 5.0	31.5 ± 5.4	25.8 ± 6.2	24.8 ± 5.8

TABLE 6.19   ANOVA Table: Swim Speed and Water Temperature (°C)

	df	Sum of squares	Mean square	F value	p value
Subject	5	567	113		
Water temperature	3	265	88	19.603	<0.0005
Water temperature * subject	15	68	4.5		

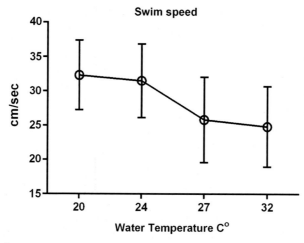

FIGURE 6.4   There appears to be a change in swim speed as a function of water temperature. The warmer the temperature the slower the swim speed. Points indicate group means and the error bars are ± SD.

different water temperatures were chosen to be compared. Using the Holm–Bonferroni Šidák MCP test we see that there is a progressive decline in swim speeds. As the temperature of the water increases the overall swim speeds decline as evidenced by the fact that both the 20 and 24 °C temperatures are significantly faster than both the 27 and 32 °C temperatures ($p < 0.05$), while swim speeds for the latter two temperatures were not significantly different from each other ($p > 0.05$).

These data could be graphed as shown in Figure 6.4. Note that there is considerable variance within each of the temperature groups, but the magnitude of the variance appears to be about equal. This consideration about variance is especially important when using a parametric repeated measure ANOVA.

SPHERICITY

The concept of sphericity (ε) (also called circularity) applies to repeated measure ANOVAs. With a regular or between-groups ANOVA, one assumes that the different groups are **independent** and that there is **homogeneity of variance**. These two conditions are vital to the accuracy of a regular one-way ANOVA statistic. With a repeated measures ANOVA, the values obtained for the dependent variable are not independent because the same subject is used, and consequently one should expect there to be some type of "consistency" in the values for a given subject. Thus, one would expect (or assume) there to be homogeneity of variance, but in fact that is often not the case. Sphericity is the condition where there actually is equality between the variances of the differences of all the possible combinations of the different levels of the factor under consideration. If there is a violation of sphericity, then the ANOVA has a good chance of becoming more liberal and prone to making a Type I error (supporting the H_A). It is important to determine if a set of data violates the assumption of sphericity before interpreting the results.

Mauchly's Test of Sphericity

The Mauchly's test was developed by John Mauchly in 1940 and is now used extensively in all modern statistical packages when evaluating repeated measures ANOVA. This test first calculates the differences in pairs of scores in all combinations of the different treatment levels. The sign of the differences is very important. The variance of the difference of each combination is then calculated. As can be observed in Table 6.20, there is a large difference in the variance between A and C compared to B and C. The Mauchly's test would say that there is a significant difference in the variances.

When the sphericity assumption is violated, the F scores computed are not necessarily reliable and the results are more prone to concluding that the overall F score is significant. Several different correction factors have been proposed to calculate the correct level of significance. One of the most common is called the **Greenhouse–Geisser estimate**[26] (sometimes also referred to as the Geisser–Greenhouse correction) that adjusts the df. The results of the Mauchly's test should be nonsignificant if the conditions of sphericity have been met.

Here is an example of how it might make a difference. In a recent study, an investigator was testing whether or not a number of different "natural" compounds could enhance word list recall in a group of individuals. Each subject was tested on a different day after consuming one of four different

TABLE 6.20   Example of Mauchly's Test

Treatment A	Treatment B	Treatment C	A−B	A−C	B−C
15	20	14	−5	1	6
24	28	20	−4	4	8
32	51	34	−19	−2	17
31	52	48	−21	−17	4
61	57	34	−6	17	23
	Variance		69	149	65

natural compounds 1 h before testing. All of these natural compounds have been glorified in the popular press as the next great memory aid. Each subject was allowed to study a word list for 60 s and then tested 5 min later for their ability to recall words on the list. A perfect score was 16 correct words recalled. Table 6.21 shows the average number correct words over 10 trials for each subject.

A standard one-way repeated measure ANOVA would produce Table 6.22.

Subsequent post hoc testing would show that the HCX compound had superior effects on word list recall. However, in this study, the Mauchly's test shows that the sphericity assumption was violated and the Geisser–Greenhouse correction factor needed to be applied. The new output would be that shown in Table 6.23.

Because the F value is now **not significant** at the $p < 0.05$ with this correction factor, there is no subsequent post hoc testing, and the H_o is not rejected. The conclusion is that in this experiment, the natural compounds did not differ for word list memory recall.

TABLE 6.21   Natural Compounds and List Recall

Subject No.	SalO	Crnb	HCX	KrO
1	7.2	10.2	8.4	7.5
2	8.3	7.9	14.1	6.8
3	6.1	8.0	9.4	8.1
4	7.3	10.5	12.2	8.2
5	11.5	6.6	10.1	10.2
6	9.3	7.2	9.3	7.2
7	10.7	8.8	11.2	9.3
8	8.1	11.2	10.1	7.2
Mean ± SD	8.6 ± 1.8	8.8 ± 1.7	10.6 ± 1.8	8.1 ± 1.2

TABLE 6.22   ANOVA Table: Natural Compounds and List Recall

	df	Sum of squares	Mean square	F value	p value
Subject	7	16	2.3		
Natural compound	3	29.4	9.8	3.414	0.0362
Natural compound * subject	21	60.2	2.9		

TABLE 6.23   ANOVA Table: Natural Compounds and List Recall with Correction Factor

	df	Sum of squares	Mean square	F value	p value
Subject	7	16	2.3		
Natural compound	3	29.4	9.8	3.414	0.0556
Natural compound * subject	21	60.2	2.9		

The results for this analysis would be reported as: "A one-way repeated measure ANOVA was applied to the data to evaluate the effect for natural compounds on word list learning. The Mauchly's test of sphericity showed a significant effect and the Geisser–Greenhouse correction factor was applied. The results failed to show a significant difference between the natural compounds [$F(3,21) = 3.414$, $p > 0.05$]."

SUMMARY

- If you are only comparing two groups use a simple t-test.
- If you are performing multiple t-tests then use a correction factor.
- A one-way ANOVA is ideal for testing for differences with multiple levels of a single variable.
- An ANOVA can only tell you if there is a difference between group means but not which one or ones are different.
- Post hoc tests are used to compare different group means following an ANOVA.
- Not all post hoc comparisons are created equal and many have important restrictions that limit their appropriateness.
- It is totally inappropriate to test your data with multiple different post hoc tests to find the one that you "feel" best explains your hypothesis.

References

1. Winer BJ, Brown DR, Michels KM. Statistical Principles in Experimental Design. 3rd ed. Boston: McGraw Hill; 1991.
2. Bartlett MS. Properties of sufficiency and statistical tests. Proc Royal Stat Soc. 1937;160: 268–282.
3. Brown MB, Forsythe AB. The ANOVA and multiple comparisons for data with heterogeneous variance. Biometrics. 1974;30:719–724.
4. Levene H. Robust test for equality of variances. In: Olkin I, Ghurye SG, Hoeffding W, Madow WB, eds. Contributions to Probability and Statistics: Essays in Honor of Harold Hotelling. Stanford: Stanford University Press; 1960:278–292.

5. O'Brien RG. A simple test for variance effects in experimental designs. *Psych Bull.* 1981; 89:570−574.
6. Hartly HO. Testing the homogeneity of a set of variances. *Biometrika.* 1940;31:249−255.
7. Hartly HO. The maximum F-ratio as a short-cut test for heterogeneity of variance. *Biometrika.* 1950;37:308−312.
8. Kirk RE. *Experimental Design.* London: SAGE Publications; 2013.
9. Dunn OJ. Multiple comparisons among means. *J Amer Stat Assoc.* 1961;56:52−64.
10. Šidák Z. Rectangular confidence regions for the means of multivariate normal distributions. *J Amer Stat Assoc.* 1967;62:626−633.
11. Holm S. A simple sequentially rejective multiple test procedure. *Scan J Stat.* 1979;6: 65−70.
12. Hockberg Y. A sharper Bonferroni procedure for multiple tests of significance. *Biometrika.* 1988;75:800−802.
13. Fisher RA. *The Design of Experiments.* Edinburgh: Oliver & Boyd, Ltd; 1949.
14. Hayter AJ. The maximum familywise error rate of the Fisher's least significant difference test. *J Amer Stat Assoc.* 1986;81:1000−1004.
15. Tukey J. *The problem of multiple comparisons: unpublished dissertation.* Princeton, N.J: Princeton University; 1953.
16. Kramer CY. Extension of multiple range tests to group means with unequal numbers of replications. *Biometrics.* 1956;12:7−10.
17. Newman D. The distribution of range of samples from a normal population expressed in terms of an independent estimate of the standard deviation. *Biometrika.* 1939;31:20−30.
18. Keuls M. The use of the "studentized range" in connection with an analysis of variance. *Euphytica.* 1952;1:112−122.
19. Duncan DB. Multiple range and multiple F tests. *Biometrics.* 1955;11:1−42.
20. Games PA, Howell JF. Pairwise multiple comparison procedures with unequal n's and/ or variance. *J Ed Stat.* 1976;1:113−125.
21. Tamhane A. A comparison of procedures for multiple comparisons of means with unequal variances. *J Amer Stat Assoc.* 1979;74:471−480.
22. Scheffé H. A method for judging all contrasts in the analysis of variance. *Biometrika.* 1953;40:87−104.
23. Dunnett CW. A multiple comparison procedure for comparing several treatments with a control. *J Amer Stat Assoc.* 1955;75:1096−1121.
24. Dunnett CW. New tables for multiple comparisons with a control. *Biometrics.* 1964;20: 482−491.
25. Dunnett CW. Pairwise multiple comparisons in the unequal variance case. *J Amer Stat Assoc.* 1980;75.
26. Greehouse SW, Geisser S. On methods in the analysis of profile data. *Psychometrika.* 1959; 24:95−112.

7

Two-Way Analysis of Variance

Stephen W. Scheff

**University of Kentucky Sanders–Brown Center on Aging,
Lexington, KY, USA**

One-way ANOVAs are always jealous of two-way ANOVAs because they can have interactions with their variables. **Anonymous**

Up to this point the discussion has really centered on the simplest form of an analysis of variance (ANOVA), the one-way ANOVA, where you have multiple levels of a single factor (e.g., Antioxidant). There are always experimental designs that want to look at outcome after manipulating two or more different factors. If there are only two independent variables (with multiple levels) being manipulated, then this type of ANOVA is called a **two-way ANOVA**. If there are three or more factors (with different levels) being evaluated, then it is called a **multifactorial ANOVA**. Often times a biostatistician will call it a **two-way factorial design**, but it is still just a two-way ANOVA. When only a single dependent variable is quantified, the analysis is termed a **univariate analysis**. Naturally, if two dependent variables are analyzed simultaneously (e.g., cell firing, cell size), then it is called a **bivariate analysis**. So if the biostatistician says you have a univariate two-way factorial design, they are simply saying that you are using a two-way ANOVA and evaluating a single outcome measure.

Let us consider the following experiment in which four different levels of antioxidants (Factor 1) are tested on two different types of mice (Factor 2), wild type (Wt) and transgenic-B7 (TrgB7) (Table 7.1). The outcome measure is cell firing rate. It is a balanced design because there are equal numbers of subjects for each condition. The question one might ask is whether or not antioxidants differentially affect the cell firing rate in TrgB7 mice. To analyze this, you would use a 4 × 2 factorial ANOVA, which corresponds to the different levels of the two independent variables.

Statisticians sometimes call independent variables that the experimenter manipulates **active variables**. In this case, the experimenter is applying different types of antioxidants. The other type of variable is sometimes called an **attribute variable**. In many experiments this could be gender (male, female) or race (Hispanic, white non-Hispanic, Asian, etc.). In the above experiment, the type of mouse (Wt, TrgB7t) is an

TABLE 7.1 Two-Way ANOVA: Antioxidant- and Mouse-Type Cell Firing

	Antiox A		Antiox B		Antiox C		Antiox D	
	Wt	**TrgB7**	**Wt**	**TrgB7**	**Wt**	**TrgB7**	**Wt**	**TrgB7**
1	132	142	141	156	118	128	108	118
2	128	151	139	148	121	131	114	123
3	137	137	142	146	111	122	104	129
4	130	146	127	157	132	143	113	122
5	142	147	134	139	127	133	104	123
6	121	131	129	142	115	122	100	121
7	134	140	143	155	126	120	107	125
8	126	139	152	147	119	136	103	131

attribute variable. As with the one-way ANOVA, the dependent variable (e.g., cell firing) must be a continuous variable.

A very common question that is asked: why use a multifactorial ANOVA instead of using two separate one-way ANOVAs? In the above experiment one could first run a one-way ANOVA on the variable "antioxidants" and then a second one-way ANOVA on "mouse type." While this is quite true, what one would not be able to test is the possible **interaction** between the two independent variables. One of the questions we wanted to answer was whether or not any of the antioxidants affect the transgenic animals differently from the wild-type animals. An added feature of the two-way (multifactorial) ANOVA is that it provides information about the effects of the two different independent variables separately. In essence, it performs the separate one-way ANOVAs automatically.

CONCEPT OF INTERACTION

A major concept in discussing a two-way ANOVA is the possibility of an interaction among the different independent variables. In order to have an interaction there must be at least two different independent variables. As an example we will use gender (male/female) and mouse type (wild type/transgenic) for evaluating the number of GFP cells in the ventral forebrain. This would involve a simple 2 × 2 table (Table 7.2).

The mean number of cells for the male Wt mice is 260 and for the Trg 125. For the female Wt mice the mean is 180 and for the female Trg it is 495. Figure 7.1 is a simple plot to show the **interaction** for the data in Table 7.2.

In this particular example, there is a large difference between the female Trg and the Wt mice compared to the male mice. From Figure 7.1 it is obvious that cell counts dramatically increase in the female Trg mice and appear to decline in the male mice. This is an interaction. An interaction exists when there is a differential affect across one of the independent variables. **If the lines are parallel then there is no interaction**. By convention, the interaction is discussed in relation to whatever variable is plotted on the X-axis. Here it would be the gender of the mice.

TABLE 7.2   Cell Counts in Ventral Forebrain

	Wild type (Wt)	Transgenic (Trg)
Male	260	125
Female	180	495

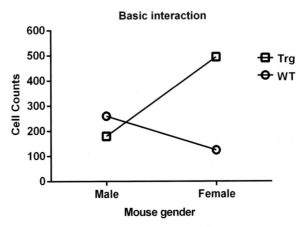

FIGURE 7.1   This graph shows a basic interaction that is indicated because the lines are not parallel. The female Trg mice respond very differently compared to the female Wt mice.

We could certainly run a simple *t*-test looking at the differences between the Wt and the Trg mice groups. The mean of the Wt would be 200 and the Trg 310. One would say that the Trg mice have significantly more GFP neurons than the Wt. But of course this would be a most inappropriate conclusion because the primary effect of the Trg was only seen in the female mice. You would also have to say that the Wt mice had higher values if the gender was male. Because there is an interaction, one must use extreme caution when talking about main effects. The main effects here are the gender of the mice and the mouse type. Some statisticians would say if you have a significant interaction you "cannot" talk about main effects.

A common misconception is that you only have an obvious interaction when you graph the data and the lines actually cross, such as in the example above. However, all the lines have to do is "significantly" **deviate** from not being parallel.

In both of the graphs in Figure 7.2, there is a significant interaction based on gender. The difference between the male Wt and the Trg values is significantly different from the difference between the female Wt and the Trg values. Because these lines do not actually cross, they are called ordinal, and when they do cross, such as in Figure 7.1, they are called **disordinal**.

The interpretation of an interaction becomes more complicated as the levels of one factor increases. For instance, instead of Wt and Trg mice, the investigator investigated three different strains of mice (C57Bl6, BALB/c, and ICR). In this example (Figure 7.3) there are six different groups that are plotted but only two different factors (gender, mouse strain). Because there is an interaction, the investigator cannot discuss main effects.

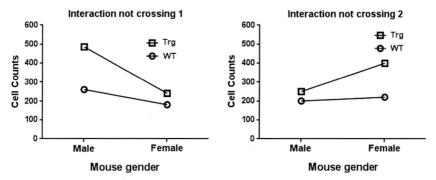

FIGURE 7.2   Even though the two lines do not cross in either of the graphs there is a strong interaction because the lines are not parallel. If the lines had been extended they would deviate.

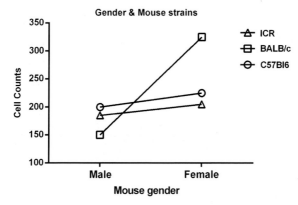

FIGURE 7.3   This interaction involves three different groups. While there does not appear to be an interaction involving the BALB/c and C57Bl6 groups, the female ICR mice are very different from the other female mice.

However, the investigator can certainly run a post hoc comparison and discuss each of the six groups separately.

DIFFERENCE BETWEEN ONE-WAY AND TWO-WAY ANALYSIS OF VARIANCE

A two-way ANOVA is a parametric statistic and thus it follows all the same requirements that the one-way ANOVA does. Each of the samples should be independent and randomly chosen from the population of interest. The populations should be normally distributed. There should be homogeneity of variance (Chapter 6). Although it is not necessary, the two-way ANOVA works best if there are the same number of subjects per treatment group (balanced design), but it still works with uneven numbers/group. Just like the one-way ANOVA, the two-way uses the F distribution. Instead of there being a single null hypothesis, with a two-way ANOVA there are actually three different H_o.

For the above experiment evaluating antioxidants and type of mouse on cell firing rates (Table 7.1), the different null hypotheses are:

1. There is no difference between the different antioxidants in terms of changing cell firing.
   a. H_A says that there is at least one antioxidant that changes cell firing different from the others.
2. There is no difference between the wild-type and transgenic animals in terms of changing cell firing.
   b. H_A says that the wild-type and transgenic animals have different cell firing rates.
3. There is no interaction between the antioxidants and the type of mouse.
   c. H_A says there is at least one antioxidant that will affect one group of mice differently in terms of cell firing.

Here is another way of thinking about the difference between a one-way ANOVA and a two-way ANOVA. In the one-way ANOVA such as in Table 6.4, the analysis considered a total of <u>four different group means</u> and these can be placed into <u>four different cells</u> in a table (Table 7.3).

In the two-way ANOVA (Table 7.1), because of the addition of the second variable (type of mouse), there are eight different group means used in the analysis, placed into <u>eight different cells</u> (Table 7.4). The analysis considers all eight of the cells but also considers the values in the rows and columns separately.

TABLE 7.3 Cells Used in One-Way ANOVA

Antioxidant A Mean	Antioxidant B Mean	Antioxidant C Mean	Antioxidant D Mean

TABLE 7.4 Cells Used in Two-Way ANOVA

Wt Antioxidant A Mean	Wt Antioxidant B Mean	Wt Antioxidant C Mean	Wt Antioxidant D Mean
TrgB7 antioxidant A mean	TrgB7 antioxidant B mean	TrgB7 antioxidant C mean	TrgB7 antioxidant D mean

TABLE 7.5   ANOVA Table for Cell Firing Effect of Antioxidant and Mouse Type

	df	Sum of squares	Mean square	F value	p value	Power
Antioxidants	3	7417.4	2472.5	57.087	<0.0001	1.000
Mouse type	1	2150.6	2150.6	49.657	<0.0001	1.000
Antioxidant × Mouse	3	190.3	63.4	1.465	0.234	0.357
Residual	56	2425.4				
Total	63	12183.6				

TABLE 7.6   Means Table for Cell Firing: Effect of Antioxidant and Mouse Type

Groups	N	Mean	SD
Antiox A Wt	8	131.3	6.6
Antiox B Wt	8	138.4	8.1
Antiox C Wt	8	121.1	6.9
Antiox D Wt	8	106.6	4.9
Antiox A TrgB7	8	141.6	6.3
Antiox B TrgB7	8	148.8	6.7
Antiox C TrgB7	8	129.4	8.0
Antiox D TrgB7	8	124.0	4.2

Table 7.5 is a typical output of the data in Table 7.1 using one of the commercial statistical packages. Many of these packages include a "means" table such as that shown in Table 7.6.

INTERPRETING A TWO-WAY ANALYSIS OF VARIANCE (WHAT DO THESE RESULTS ACTUALLY TELL US?)

As shown in Table 7.5, there was a significant main effect between the different antioxidants used. The analysis collapsed all the data between the two different mouse types, and simply evaluated whether or not at least one antioxidant mean was different from the other means. There was a large F score (57.087) with significance indicating that there was less than 0.0001% chance that the observed results were due to random sampling, supporting the H_A and rejecting the H_0. This means that at least one of the four antioxidants tested was significantly different. However, we do not know which one because that takes additional testing, but in a manuscript one can say there was a significant main effect for antioxidant.

By using the ANOVA table, one can calculate what fraction of the total variance results from the different antioxidants used. One simply divides the sum of squares of the antioxidant factor (7417) by the total sum of squares (12183) and the result is approximately 61%. That means that approximately 61% of the variance in the different group means can be explained by the type of antioxidant.

There was also a significant main effect for mouse type ($F = 49.657$), indicating that one of the two groups had a higher or lower cell firing rate when compared to the other. What the analysis did was collapse all the data across the different antioxidant types and essentially just compared the group means of the two mouse types. We can visually compare the two and it appears that the TrgB7 group had a higher firing rate than the Wt cohort. These data then reject the H_0 and support the H_A. There is less than a 0.0001% chance that the observed results were the result of random sampling. The mouse type accounts for approximately 18% of the variance.

The analysis also revealed that there was **no interaction** between the antioxidant and the mouse type. In the table, the row labeled **Antioxidant × Mouse**, shows the p value to be >0.2, which means that both groups of mice respond in a statistically identical fashion to each of the antioxidants. If these data are graphed only to look at the possible interaction, it would look like Figure 7.4.

While there appears to be a difference between the Wt and the TrgB7 cohort for each of the antioxidants used, the direction of the change is the same even with antioxidant D. Using post hoc comparisons for the main effect for antioxidant, there are multiple significant differences between groups (e.g., A is significantly different from C and D). However, one must remember that the data between mouse types are combined for this type of comparison. Because there was no significant interaction, comparisons between many different groups (e.g., antioxidant D-treated Wt versus antioxidant D-treated TrgB7) are **not allowed**. One can only discuss overall main effect and the graphs in Figure 7.5 would be seen.

In the text of a manuscript the results might read: "A two-way ANOVA (Antioxidant × Mouse Type) was performed and the results indicated a significant main effect for antioxidant [$F\,(3,56) = 57.087$, $p < 0.0001$] and also for mouse type [$F\,(1,56) = 49.657$, $p < 0.0001$]. There was no significant interaction between these two variables [$F\,(3,56) = 1.465$, $p > 0.2$]." One would then go on to discuss the individual post hoc testing for antioxidant comparisons. It is extremely important to report the degrees of freedom because it allows the reader to understand exactly

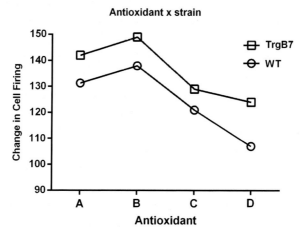

FIGURE 7.4   Although it appears that there is deviation comparing the TrgB7 and Wt mice with antioxidant D, the separation was not strong enough to warrant a significant difference.

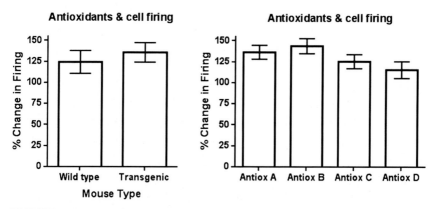

FIGURE 7.5   These two graphs show the results of the analysis of just the main effects since there was no interaction. Bars represent group means ± SD.

how many groups were used in the analysis. Reporting the actual F value allows the reader to appreciate the robustness of the main effects.

How would things change if in fact the data were just a little different and there appeared to be an interaction between antioxidant and mouse type? In this data set (Table 7.7) the response of the TrgB7 mice is a little greater to antioxidant D. With these data there would be no need to actually talk about main effects because the response to each variable is regulated by the other variable. Here is the altered data set with the analysis output indicating a significant interaction. The statistical analyses of these data are shown in Table 7.8.

TABLE 7.7   Two-Way ANOVA: Antioxidants and Mouse Type and Cell Firing

	Antiox A		Antiox B		Antiox C		Antiox D	
	Wt	TrgB7	Wt	TrgB7	Wt	TrgB7	Wt	TrgB7
1	132	142	141	156	118	128	108	125
2	128	151	139	148	121	131	114	133
3	137	137	142	146	111	122	104	129
4	130	146	127	157	132	143	113	122
5	142	147	134	139	127	133	104	134
6	121	131	129	142	115	122	100	129
7	134	140	143	155	126	120	107	125
8	126	139	152	147	119	136	103	131

TABLE 7.8   ANOVA Table for Cell Firing: Effect of Antioxidant and Mouse Type

	df	Sum of squares	Mean square	F value	p value	Power
Antioxidants	3	6410.5	2136.8	49.379	<0.0001	1.000
Mouse type	1	2588.3	2588.3	59.810	<0.0001	1.000
Antioxidant × Mouse	3	459.2	153.1	3.537	**0.0203**	0.758
Residual	56	2423.4	43.3			
Total	63	11881				

The most interesting part of the analysis is whether or not there is an interaction between the two independent variables. The fact that the interaction had a significant p value ($p < 0.03$) is important. What that means is that at least one of the antioxidants affected either the Wt or the TrgB7 mouse group differently from the rest. There is less than a 3% chance that the results were due to simple random sampling. This interaction accounts for approximately 4% of the variance, but of course we do not know which of the groups are different, and this requires additional analysis or simply graphing the results.

By graphing the main variables, just for the purpose of showing possible interactions, the group that drives the interaction often becomes quite obvious (Figure 7.6). What does an interaction actually mean? For the interaction, the two-way analysis of variance does not really consider the main effect means, but rather considers the individual cell means. In our example above, the null hypothesis is actually stating that the

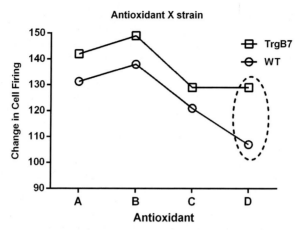

FIGURE 7.6   These data are very similar to those graphed in Figure 7.4 with the exception that Antioxidant D values for the TrgB7 mice have been altered to show an interaction.

antioxidants will result in similar firing patterns for both types of mice. What we find is that antioxidants A and B increased the firing rate for both types of mice, and antioxidant C shows a similar decline for both mouse types compared to both A and B. But antioxidant D shows a differential response for the two mouse groups. The analysis says there is a significant ($p < 0.03$) interaction and this is where it is. What you can say is that the effect of the different antioxidants (here specifically antioxidant D) is modified or qualified by the interaction with the mouse types.

In the manuscript you would not actually graph the interaction but instead graph the different group means and simply describe the interaction. The group means table provided by the statistical software is shown in Table 7.9 and the graph of these data in Figure 7.7.

There are several ways to report this in a manuscript. Here is an example for the analysis of the data in Table 7.7: "A 4×2 ANOVA revealed a significant interaction between antioxidants and mouse type [$F(3,56) = 3.54$, $p < 0.03$]." This would be followed by the results of the post hoc testing. Depending on the post hoc test used, there appears to be a significant increase in cell firing ($p < 0.05$) in the transgenic mice following all antioxidants except C. The largest difference between the mouse strains was observed with antioxidant D. Depending on the original hypothesis, one could further elucidate comparisons such as the fact that in the wild-type animals antioxidants A and B significantly increased the firing rate compared to C and D. This was also the case for the transgenic mice.

The same rules used for multiple comparisons following a one-way ANOVA (Chapter 6) also apply for a two-way analysis of variance.

TABLE 7.9   Means Table for Cell Firing: Effect of
Antioxidant and Mouse Type

Groups	N	Mean	SD
Antiox A Wt	8	131.3	6.6
Antiox B Wt	8	138.4	8.1
Antiox C Wt	8	121.1	6.9
Antiox D Wt	8	106.6	4.9
Antiox A TrgB7	8	141.6	6.3
Antiox B TrgB7	8	148.8	6.7
Antiox C TrgB7	8	129.4	8.0
Antiox D TrgB7	8	128.5	4.2

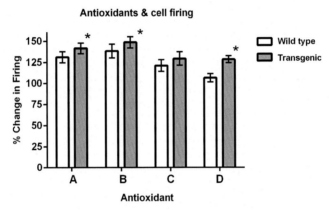

FIGURE 7.7   The graph shows the group means $\pm$ SD. Note that the * indicates $p < 0.05$ compared to the wild-type mouse receiving the same antioxidant treatment.

TWO-WAY REPEATED MEASURE ANALYSIS OF VARIANCE

Both Factors Repeated

The major difference between the basic design of this type of study and the one presented in the previous chapter is the fact that there are now two independent variables being manipulated, with the same subjects being assessed for each of the variables. As an example, the previously described experiment (Table 6.7), featuring different natural compounds and memory recall, can be made more complicated by also studying the way the word lists are presented. Instead of being allowed to study (**S**) the word list for 60 s, the individual listens (**L**) to the word list repeated twice over a 60 s period. The two variables are (1) natural compounds and (2) word list presentation. Table 7.10 shows the individual subject means for 10 trials under each of the conditions. In this study the <u>same subjects</u> are used for both word presentations and for each of the natural compounds. There are a total of eight different individuals tested. A typical printout for this type of analysis is shown in Table 7.11.

In order to interpret these results one must first address the main effects. Did the natural compounds have an influence on the word list learning? What this type of analysis does is collapse all the scores across presentation type, and essentially compares only the four different natural compound means. However, instead of only using eight values for each compound like in the one-way ANOVA, it is now using 16 values, eight from the studied list and eight from the listened list for each subject. The analysis shows a very robust effect with a very large F score (30.961) that is significant at the $p < 0.0001$ level. Thus one would report that the natural compounds altered word list learning. Figure 7.8 shows a graph of the main effect for natural compounds. By using a post hoc MCP test such as the Holm–Šidák we find that the HCX group is significantly different from the other three compounds, which are not different from each other.

There is also a significant main effect for type of presentation of the word lists with an F score of 37.252 which garners an alpha at the 0.0005 level. What this part of the analysis does is collapse across all the different compounds, and tests whether or not there was a difference between the studying and the listening of the word list. By simply looking at the table of mean values, since there were only two levels of word list presentation, it is easy to recognize that the "Studied" type of presentation was better. Figure 7.9 is a graph of the means table. If there were more than two levels, then a post hoc MCP would be required to detect the differences.

It is easy to see in Figure 7.10 that studying the list resulted in higher word recall scores and that the HCX natural compound was superior compared to the other agents tested. Close observation reveals that both learning conditions responded the same to each compound. Since there was no interaction [$F(3,21) = 0.781, p > 0.1$] (Figure 7.10), no other type of comparison is warranted.

This might be reported in the literature as follows: A two-way repeated measure ANOVA (Natural Compounds by Presentation Style), with both factors repeated, was applied to the data. There was a significant main effect for natural compound [$F(3,21) = 30.96, p < 0.0001$] and also the presentation style [$F(1,21) = 37.25, p < 0.0005$]. There was no significant interaction ($p > 0.1$). Overall, word recall was significantly better when the word list was studied as opposed to spoken. Post hoc comparison of

TABLE 7.10   Two-Way Repeated Measure ANOVA: Antioxidants, Presentation, Word Recall

Subject	SalO-S	SalO-L	Crnb-S	Crnb-L	HCX-S	HCX-L	KrlO-S	KrlO-L
1	9.4	7.8	9.8	8.4	12.2	9.6	7.2	6.8
2	8.8	7.2	10.0	8.2	11.6	10.9	7.8	7.2
3	9.0	8.0	7.6	7.6	9.6	9.8	8.6	7.8
4	9.4	7.6	8.8	7.4	9.8	9.4	8.8	6.4
5	8.4	7.0	7.8	7.0	11.8	10.1	7.6	7.0
6	7.0	6.8	8.6	7.6	10.4	10.4	9.6	7.8
7	7.8	7.4	9.2	8.0	10.0	10.0	7.6	7.2
8	8.2	7.6	9.0	7.8	10.6	10.6	7.2	6.8
Mean ± SD	8.5 ± .8	7.8 ± .5	8.9 ± .9	7.8 ± .5	10.8 ± .8	10.1 ± .5	8.1 ± .9	7.2 ± .4

TABLE 7.11   Repeated Measure ANOVA: Antioxidants, Presentation, Word Recall

	df	Sum of squares	Mean square	F value	p value
Subject	7	2.93	0.419		
Natural compound	3	76.8	25.6	30.961	<0.0001
Presentation style	1	13.7	13.4	37.252	<0.0005
Compound * Presentation	3	0.53	0.18	0.781	>0.1
Comp * Pres * Subject	21	4.70	0.22		

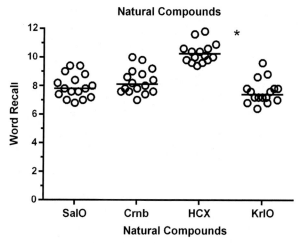

FIGURE 7.8   Since there is no interaction, one can only present the results of the main effect. Here, the data from both presentation types (Study, Listen) have been collapsed together such that each group represents 16 scores. The data are presented as a scatterplot to reinforce the idea that the data from the different presentation styles have been collapsed together. Bars represent group median *$p < 0.05$.

the different natural compounds revealed that the HCX compound resulted in significantly greater word recall compared to the other natural compounds tested ($p < 0.01$), which were not significantly different from each other.

How would the <u>analysis</u> and interpretation change **if there was a significant interaction?** For example, if the word recall scores were higher for the HCX group that listened to the word list (Table 7.12), then there would be a significant interaction as indicated in Table 7.13.

In this analysis we see a significant interaction between the presentation type and the natural compounds but it was not extremely strong

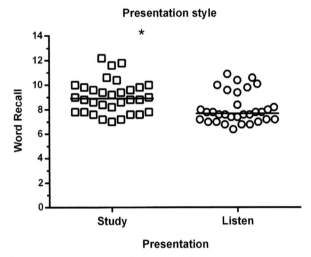

FIGURE 7.9   Since there is no interaction, one can only present the results of the main effect. Here, the data from different natural compounds (SalO, Crnb, HCX, Krl) have been collapsed together such that each group represents 32 scores. The data are presented as a scatterplot to reinforce the idea that the data from the different natural compounds have been collapsed together. Normally these type of graphs are shown as bar graphs. The horizontal bars represent group medians *$p < 0.05$.

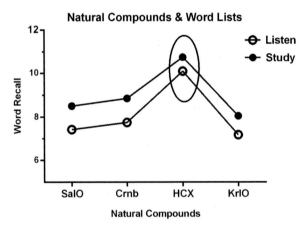

FIGURE 7.10   You might plot the group data to help understand the possible shift in word recall scores but this type of graph is not published. The natural compounds appear to have had the same influence on the different presentation styles.

($p < 0.03$), but nevertheless an interaction. In this case, the interaction (as indicated by the circle in Figure 7.11) showed no difference ($p > 0.05$) between the recall scores for the presentation style in the HCX group, whereas the other comparisons were significantly different. This could be

TABLE 7.12   Two-Way Repeated Measure ANOVA: Two Factors Repeated

Subject	SalO-S	SalO-L	Crmb-S	Crmb-L	HCX-S	HCX-L	KrlO-S	KrlO-L
1	9.4	7.8	9.8	8.4	12.2	11.6	7.2	6.8
2	8.8	7.2	10.0	8.2	11.6	11.2	7.8	7.2
3	9.0	8.0	7.6	7.6	9.6	10.2	8.6	7.8
4	9.4	7.6	8.8	7.4	9.8	9.4	8.8	6.4
5	8.4	7.0	7.8	7.0	11.8	11.2	7.6	7.0
6	7.0	6.8	8.6	7.6	10.4	9.8	9.6	7.8
7	7.8	7.4	9.2	8.0	10.0	9.4	7.6	7.2
8	8.2	7.6	9.0	7.8	10.6	10.6	7.2	6.8
Mean ± SD	8.5 ± .8	7.8 ± .5	8.9 ± .9	7.8 ± .5	10.8 ± .8	10.4 ± .9	8.1 ± .9	7.2 ± .4

TABLE 7.13   ANOVA Table: Two-Way Repeated Measure ANOVA

	df	Sum of squares	Mean square	F value	p value
Subject	7	4.82	0.688		
Natural compound	3	86.7	28.9	28.039	<0.0001
Presentation style	1	11.39	11.4	48.489	<0.0002
Compound * Presentation	3	1.6	0.52	3.909	< 0.03
Comp * Pres * Subject	21	2.6	0.12		

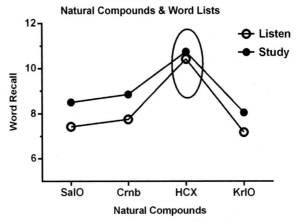

FIGURE 7.11   These data are almost exactly the same as those in Figure 7.10 with the exception that now there is an interaction with the presentation style and the HCX compound. There data lines are no longer parallel at this juncture.

interpreted as saying the HCX natural compound had a greater effect regardless of whether the word list was studied or spoken to the subject. In the results section the post hoc testing will be very important and should reflect the various comparisons.

Because of this interaction, many statisticians would say that one should not talk about main effects and instead the researcher should concentrate on individual group comparisons. This would actually be too conservative. What these data clearly indicate is that there is a significant overall difference between natural compounds. With the possible exception of the HCX group, there was also a significant effect of the way the word list was presented. Given the fact that the interaction was not very robust, it would be wise to discuss these differences in the reporting of results.

TABLE 7.14   Two-Way Repeated Measure ANOVA: One Factor Repeated

Subject	SalO-S	SalO-L	Crmb-S	Crmb-L	HCX-S	HCX-L	KrlO-S	KrlO-L
1	9.4		9.8		12.2		7.2	
2	8.8		10.0		11.6		7.8	
3	9.0		7.6		9.6		8.6	
4	9.4		8.8		9.8		8.8	
5	8.4		7.8		11.8		7.6	
6	7.0		8.6		10.4		9.6	
7	7.8		9.2		10.0		7.6	
8	8.2		9.0		10.6		7.2	
9		7.8		8.4		11.6		6.8
10		7.2		8.2		11.2		7.2
11		8.0		7.6		10.2		7.8
12		7.6		7.4		9.4		6.4
13		7.0		7.0		11.2		7.0
14		6.8		7.6		9.8		7.8
15		7.4		8.0		9.4		7.2
16		7.6		7.8		10.6		6.8
Mean ± SD	8.5 ± .8	7.8 ± .5	8.9 ± .9	7.8 ± .5	10.8 ± .8	10.4 ± .9	8.1 ± .9	7.2 ± .4

In both of these examples, the Mauchly's test (Chapter 6) did not indicate a violation of sphericity. Consequently, it was not necessary to apply the Geisser–Greenhouse correction factor. The Mauchly's test only applies if you have three or more levels of one of the factors in a two-way repeated measures design. In fact, it is probably impossible to get negative sphericity in a simple 2 × 2 repeated measures design, so if that is the design you are using do not even worry about sphericity.

Single Factor Repeated

A two-way repeated measure ANOVA can be used with a different type of design. Instead of both factors repeated in each subject, only one factor is repeated. For the above experiment, it could be the case that a total of 16 subjects were used (Table 7.14). Half of the subjects were allowed to study (S) the word list for 60 s, and the other half of the subjects listened (L) to the word list repeated twice over a 60 s period. Each subject was tested with each of the natural compounds. For simplicity of this example the same data are used just to emphasize that there is a dramatic difference in the outcome of the study.

The main effects for both Natural Compound and Presentation Style are significant as in the previous analysis (Table 7.13) but there is **no interaction** to report even though the individual group means are the same. Compare the data analysis in Table 7.13 with Table 7.15. In Table 7.13 there was a significant interaction and in Table 7.15 there is no significant interaction. The reason lies in the fact that while the natural compounds were repeated across subjects, the presentation style of the word list was not. One might report the results with simple bar graphs

TABLE 7.15   ANOVA Table: Two-Way Repeated Measure with One Factor Repeated

	df	Sum of squares	Mean square	F value	p value
Subject	14	6.46	0.462		
Natural compound	3	86.7	28.9	49.680	<0.0001
Presentation style	1	11.39	11.4	24.671	<0.0002
Compound * Presentation	3	1.6	0.52	0.892	.453
Comp * Pres * Subject	42	24.4	0.582		

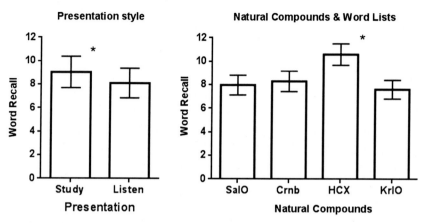

FIGURE 7.12 Because there was no significant interaction, only the main effects are presented. These results are very interesting since there was a main effect when both factors were repeated and the interaction is lost when only one factor was repeated. Bars represent group means ± SD *$p < 0.05$.

(Figure 7.12). The first would show that the presentation style had an effect and also that at least one of the natural compounds had an effect.

When reporting the results of a repeated measure two-way ANOVA it is important to state whether or not both factors are repeated. You are probably thinking, "Can't the reader tell that from the design of the experiment?" If in fact the reader wants to take the time to carefully plot out your experiment, then yes. However, many researchers will simply look at the figures and the results section and not even pay close attention to the methods, unless of course there is something that intrigues them.

SUMMARY

- A two-way ANOVA is basically the same as a one-way ANOVA.
- The major question is whether or not there is an interaction between the variables.
- The simplest way to check for an interaction is to plot the main effects.
- The same post hoc comparison rules apply to both a one-way and a two-way ANOVA.
- A two-way repeated measure ANOVA can either repeat both factors or repeat only one factor, which changes the way the analysis is done.

Nonparametric Statistics

Stephen W. Scheff

University of Kentucky Sanders—Brown Center on Aging, Lexington, KY, USA

Statistics is exciting because you get to play with others' data while telling them their research is crap. **Stephen Senn**

OUTLINE

Fundamental Statistical Principles for the Neurobiologist
http://dx.doi.org/10.1016/B978-0-12-804753-8.00008-7
157

Another name for nonparametric statistics is **"distribution-free statistical procedures."** What this means is that there are no assumptions made about the distribution of the data, unlike the parametric tests that "assume" a normal distribution. In short, there is no assumption about the variability of the data (forget about homogeneity of variance) and the form of the distribution. These statistics are designed to work with data that have a lot of variance either in one group or in several groups, or with data that are primarily ordinal and nominal.

Major advantages of nonparametric statistics:

1. Can be used with nominal, ordinal, interval, or ratio data
2. Are not restrictive about assumptions concerning distribution and variance
3. These tests are not affected by extreme outliers in the data
4. Can sometimes detect differences between groups that parametric statistics do not
5. Can be used with very small sample sizes
6. Can be used even when the data are skewed
7. Often very easy to calculate
8. Quite often they are very easy to understand

Major disadvantages of nonparametric statistics:

1. Less powerful than parametric statistics
   a. Parametric statistics are more powerful only when the assumptions underlying their use are valid (e.g., homogeneity of variance)
2. Can only be used with relatively "simple" experimental designs
3. Do not take advantage of all the information about a sample distribution
4. Analyze ranks rather than actual experimental values
5. Conclusions are more general because hypotheses tested are less specific

Why are nonparametric statistics less powerful than parametric procedures? The very simple reason is that the parametric statistics use all

TABLE 8.1 Rank Ordering of Numbers

Rank	1	2	3	4	5	6	7	8
	717	672	534	314	298	197	123	111

of the information and nonparametric statistics do not. For example, if you were to count the number of astrocytes in a given region of the thalamus and got the following data from eight rats:

$$197, 717, 298, 123, 672, 534, 111, 314$$

you would subsequently rank the data (for nonparametric analysis) from highest value to lowest (Table 8.1).

The distance between 534 and 314 (ranks 3 and 4) would be the same as between 123 and 111 (ranks 7 and 8). The information concerning the magnitude of the scores is lost when converted into ranks.

For some nonparametric statistics, it is assumed that no two values are the same. While this may sound rather strange, for these statistics, ties that do occur are simply eliminated from consideration. Some statistics software programs actually attempt to make a correction for ties. Still in other statistics, ties are given an average score. It is also important to remember that when reporting results involving nonparametric statistics, it is **inappropriate** to report the mean and standard deviation. Unfortunately, this mistake is made very often in journal articles. The appropriate descriptive statistic to report is the median and the range for each group either in the text or in a table.

There are so many different distribution-free procedures and individual statistical tests that it is beyond the scope of this book to present them all. The reader is advised to consult several excellent texts on the topic (Siegel and Castellan[1]; Hollanders and Wolfe[2]; Corder and Foreman[3]). Most of the commercially available statistical packages have a limited number of nonparametric statistical tests available.

SIGN TEST

This nonparametric statistic is used for ordinal data and essentially measures the relative ordering of different categories of a variable. At the end of a study evaluating the effectiveness of an antioxidant on recovery from traumatic brain injury you note a particular cluster of glial cells in the anterior thalamus contralateral to the injury that you have never observed before. It does not appear in all of the animals but you want to know if there is any evidence (at the $\alpha = 0.05$) that the antioxidant treatment is related to this "anatomical anomaly." Although you are not sure what this anatomical feature means, it might first be wise to test whether or not it is worth pursuing. The H_o is that antioxidant treatment has no relationship to the anatomical clustering of cells in the anterior thalamus. To test this you obtain histological sections from 12 animals that have had both brain injury and the antioxidant therapy. You then simply find out how many of these animals show the anatomical clustering and assign those animals a plus sign (+) and those that do not a minus sign (−). In the example in Table 8.2, there is a clustering of cells in 9 of the 12 animals. If chance alone is working, the probability of a plus is equal to the probability of a minus. Since there are only two alternatives (yes or no), $P = 0.50$. By simply consulting a binomial distribution table, we find that for n = 12, with nine pluses and $P = 0.50$, the resulting p value is 0.0737, which does not quite reach significance and thus the H_o is not rejected.

The sign test is sometimes used in a repeated measure type design. For example you might want to know if lowering the water temperature in your Morris water maze makes the rats swim faster. You record the swim speed of the subjects with the temperature at two different temperatures (27 and 24 °C). If the speed is faster at the lower temperature it is a plus and if it is slower it is a minus. If there is no difference between the two

TABLE 8.2   Sign Test

Animal	Yes	No
1	+	
2	+	
3		−
4	+	
5	+	
6	+	
7		−
8		−
9	+	
10	+	
11	+	
12	+	
Total	9	3

scores the values are ignored and the subject is not included in the analysis. Consider the data in Table 8.3.

Because animals No. 2 and No. 3 had the same value, they are not included in the analysis. In this situation the n = the total number of +'s and −'s, which is 10. In the present example the H_0 is that a lower water temperature does not increase swim speed. This is a one-tailed test because it is asserting direction. By consulting the binomial distribution table, for n = 10, with seven pluses and three minuses and $P = 0.50$, the resulting p value is 0.1719. In this case the H_0 is supported and we conclude that the lower water temperature did not significantly increase the overall swim speed. What is of importance in the sign test is the lower number, whether it is minuses or pluses. In Table 8.3 the minus signs have a lower number than the plus signs. Very often the results of a sign test are given by a simple p value such as $p > 0.1$. For somewhat large samples, some investigators will calculate a Z value. In those situations the reporting of the sign test in a journal would be the Z value (e.g., Z = 3.68) along with the p value (e.g., $p < 0.001$).

The sign test can only indicate the probability of obtaining a specific score that is in the opposite direction of the other cumulated value. It cannot indicate cause and effect. If the results are statistically significant then it indicates that the two groups were derived from different populations. What is absolutely critical is that the scores are paired if it is a repeated measure design.

TABLE 8.3   Sign Test: Water Temperature

Animal	27 °C	24 °C	+/−
1	20	23	+
2	34	34	na
3	32	32	na
4	27	39	+
5	31	33	+
6	22	20	−
7	26	29	+
8	24	28	+
9	30	32	+
10	28	31	+
11	27	26	−
12	29	28	−

WILCOXON MATCHED PAIRS SIGNED RANK TEST (WILCOXON SIGNED RANK TEST)

This particular test is also called the **Wilcoxon matched pairs test** or the **Wilcoxon signed rank test**. It is very appropriate for a repeated measure design where the same subjects are evaluated under two different conditions such as with the water maze temperature experiment in Table 8.3. It is the nonparametric equivalent of the parametric paired t-test. This is not the same as the **Wilcoxon rank sum test**, which compares two nonpaired groups and is equivalent to the parametric unpaired t-test. The Wilcoxon signed rank is more powerful than the sign test. This statistic differs from the sign test in that it considers the magnitude of the difference while the sign test does not. It uses more information from the sets of scores than the simple sign test. Because it uses more information it is considered to be more precise than the sign test. Look at the swim speed data (cm/s) in Table 8.4 and at the result of the three different statistics in Table 8.5.

If a pair of scores are equal (the same value) then they are considered tied and dropped from the analysis and the sample size is reduced. In the data below, there are two tied scores (pair No. 2 and No. 3) and three pairs of scores where the swim speed was slower in the colder water (Nos. 6, 11, 12), thus the n = 10 for this nonparametric test. What is absolutely critical in using this test is that the pairs of scores under consideration are related

TABLE 8.4   Wilcoxon Signed Rank Test

Animal	27 °C	24 °C	Difference
1	20	23	+3
2	34	34	na
3	32	32	na
4	27	39	+12
5	31	33	+2
6	22	20	−2
7	26	29	+3
8	24	28	+4
9	30	32	+2
10	28	31	+3
11	27	26	−1
12	29	28	−1

TABLE 8.5   Results: Wilcoxon Signed Rank Test

Student *t*-test	Sign test	Wilcoxon signed rank test
$p = 0.0756$	$p = 0.1719$	$p = 0.0367$

and that they are at least ordinal scale. It is unclear why this test is not used more especially in behavioral neuroscience where much of the data do not follow a normal distribution. Many statistical software programs include this statistical test.

Because of the variance in the scores, the Student *t*-test says there is no significant difference. The Wilcoxon signed rank test, which is more sensitive than the sign test, shows a very different outcome and supports the alternative hypothesis (H_A).

When To Use the Wilcoxon Test versus the Sign Test

Whenever you have data that are composed of definite scores, the Wilcoxon signed rank test is preferred. When the data are not a definite score, or if the data are observational, such as "more aggressive" versus "less aggressive" then the sign test is the appropriate statistic. Whenever there is a difference in a particular direction but the absolute quantity of that difference is not precise, and the scores are paired, then the sign test is the statistic to use.

Reporting the results in a journal article requires reporting the observed Z value, the number of observations, and the significance. Often some investigators will report the number of instances with no difference. In the above experiment, one would write: "A Wilcoxon signed rank test revealed a significant difference in the swim speeds between the two water temperatures, $n = 10$, $Z = 2.09$, $p < 0.05$. There were two pairs that showed no difference." Sometimes a T value is reported instead of the Z value. Typically the data are not graphed since it is a repeated measure analysis.

MEDIAN TEST

This is a very simple statistic that can be used to quickly determine if there is a difference between two independent samples even with unequal sample size. Like the Sign test above, it simply counts how many observations occur in a given category regardless of the magnitude of the difference.

TABLE 8.6   Median Test: Enriched Environment

No toys	Toys
19	7
16	8
16	8
20	6
8	7
9	11
7	12
11	14
10	6
10	6

Let us say that you found out that the workers in the animal care facility have been placing novel objects ("toys") in the cages of some of the rats immediately after weaning and you have been using these animals in a learning experiment. This may constitute an enriched environment and could have an effect on some of your experimental studies especially concerning recovery from brain injury. You review your laboratory notes and randomly select a group of animals that had the novel objects and 10 that did not (Table 8.6). The H_0 is that the placement of novel objects during weaning had no effect on learning. You analyze errors to criterion in the learning phase of the task. The important factor here is that the animals are randomly selected from the population. In this case, they would have to be randomly selected from the entire group of animals that had the novel objects and then from animals that did not.

What the median test does is set up a contingency table (Table 8.7) and then applies a chi-square test or X^2. In this case the chi-square test shows a value of 3.20 with a p value of 0.0736, supporting the H_0 that the addition

TABLE 8.7   Results: Median Test—Enriched Environment

Groups	<Median	>Median	Totals
No toys	3	7	10
Toys	7	3	10
Totals	10	10	20

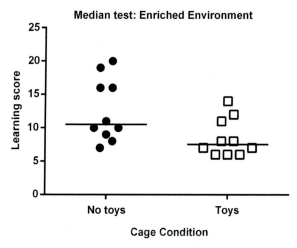

FIGURE 8.1   It is quite apparent that the two groups overlap and the medians are not that different. Each circle or square is the learning score from an individual subject.

of toys did not influence the learning. However, this particular test is not very sensitive to the Type I error, but it is a very quick and easy test to do. These results are sometimes graphed as shown in Figure 8.1.

WILCOXON RANK SUM TEST (MANN–WHITNEY U TEST)

The Wilcoxon rank sum statistic can also be used when comparing two independent groups with the added advantage that it is applicable whether the groups have equal sample size or not. This is the equivalent of the parametric unpaired *t*-test. It is sometimes also referred to as the **Wilcoxon two-sample test**, the **Wilcoxon test**, the **Wilcoxon–Mann–Whitney test**, or more commonly the **Mann–Whitney U test**. Technically the Mann–Whitney U test uses a slightly different formula but results in the same outcome. This statistic is more sensitive than the median test. As the name implies, one sums the ranking of data from a particular group and compares it to the sum of the ranks of the other group. The idea is to determine if the combined scores are randomly mixed or do the scores from one group cluster toward one end when ranked from lowest to highest. Overall this is a very simple test to run. The only stipulations are that the two samples being investigated must be independent. The variable under consideration must be continuous and the two groups have approximately equal variance. The two groups do not need to have an equal number of scores. Basically, you pool the data and rank it in ascending order, keeping track of which values belong to which group. For example, if we take the data concerning the novel objects (toys) (Table 8.6) and rank all of the data we would get Table 8.8.

Note that when three scores are identical they have identical ranking (e.g., the three 8's represent ranks 7, 8, 9 and thus each gets the rank 8). The ranks are then separated back into the two groups and one computes the U statistic. Most statistical programs will list the Mann—Whitney U test and provide a U score, a Z value, and a p value.

As can be gleaned from summary Table 8.9, the Mann—Whitney U test supports rejection of the H_0 and supports the idea that the addition of the

TABLE 8.8 Mann—Whitney U Test—Enriched Environment

Scores	Rank	No toys	Toys
6	2		2
6	2		2
6	2		2
7	5		5
7	5		5
7	5	5	
8	8	8	
8	8		8
8	8		8
9	10	10	
10	11.5	11.5	
10	11.5	11.5	
11	13.5	13.5	
11	13.5		13.5
12	15		15
14	16		16
16	17.5	17.5	
16	17.5	17.5	
19	19	19	
20	20	20	
Sum	**210**	**133.5**	**76.5**

TABLE 8.9　Results: Mann—Whitney U
Test—Enriched Environment

U	Z value	p value
21.5	−2.154	0.0312

novel objects had a significant influence on the learning. So why do the results differ between the **median test** and the **Mann—Whitney U test**? The simple answer to this is that the median test only looks at the difference between the medians of the two groups while the Mann—Whitney U looks at the difference in the shape and spread of the scores. The Mann—Whitney U is not really a test of medians. Because it can test a difference in the spread of scores, even when the medians are very similar, it is a more powerful statistic. The Mann—Whitney U test uses more information than the median test.

Reporting the results in a journal article requires reporting the U statistic (e.g., U = 31), the n's per group, and the significance. In the above experiment, one would write: "The Mann—Whitney *U* test revealed that the group exposed to toys (n = 10) was less likely to make errors in the maze acquisition than animals without toys (n = 10), U = 30, $p < 0.05$." Some statisticians suggest that as the groups become large (15—20), then the z value is reported instead of U.

The Mann—Whitney U test is really great as long as the n/group is less than 20. When the n is >20, the distribution begins to approach that of a parametric *t*-test and then the *t*-test has more power. However, if the n/group is small and there is large variance in the data then the Mann—Whitney U and *t*-test will give vastly different results. For example in a study looking at CYP-labeled neurons in the paraventricular nucleus of the hypothalamus, the following data were collected from a group of male and female rats (Table 8.10).

There is difference in the means for each group (Male, 138.7; Female, 223.3) and the question becomes: Are these two groups significantly different? If a parametric *t*-test is applied to these data one obtains the results in Table 8.11.

The *t*-statistic would support the H$_o$ that there is no difference between the groups. If these same data were analyzed using the **Mann—Whitney U test**, the opposite conclusion would be obtained, **supporting the H$_A$** that the two groups were significantly different (Table 8.12).

The reason for the difference is due to the magnitude of the difference in the variance in each of the groups. The SD for the Male group = 38.5 while the Female group = 105.5. Homogeneity of variance is an important factor in using parametric statistics such as a *t*-test. These results can be subsequently graphed as shown in Figure 8.2.

TABLE 8.10 Mann—Whitney U Test:
Neurons in Paraventricular
Nucleus

Gender	Neurons
Male	128
Male	127
Male	105
Male	214
Male	138
Female	249
Female	296
Female	137
Female	387
Female	142
Female	129

TABLE 8.11 t-Test: Neurons in Paraventricular Nucleus

Mean difference	df	t value	p value
84.7	10	1.846	0.0946

TABLE 8.12 Mann—Whitney U Test—Neurons in
Paraventricular Nucleus

U	Z value	p value
30	−2.082	0.0374

KOLMOGOROV—SMIRNOV TWO-SAMPLE TEST

The Kolmogorov—Smirnov two-sample test (K—S test) is a nonparametric statistic that tests the hypothesis that two independent samples have been drawn from the same population. It analyzes the data very differently from the Mann—Whitney U test in that rather than evaluating the location of the median it looks at a single maximum difference between the two distributions of scores. This test is often used when the question of equal variance between groups is very uncertain. It is based

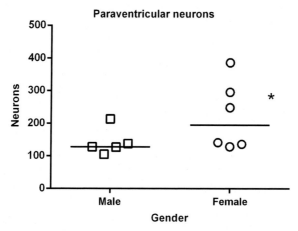

FIGURE 8.2   The Mann–Whitney U statistic is ideal when comparing two groups that have heterogeneity of variance. Note that all of the scores of the female group are above the median of the male group. Open squares and open circles represent individual subject data points. *$p < 0.05$.

on the chi-square distribution and relies on the relative expected frequencies. In the above example (Table 8.10) looking at gender-related neuron number in the paraventricular nucleus, the K–S test would report a p value of 0.14 and support the H_o. One important difference between this statistics and the Mann–Whitney U test is that the Mann–Whitney U test can handle tied scores quite well while the K–S does not. However, the K–S statistic can also be used in tests of goodness of fit. Most statisticians agree that the Mann–Whitney U is a better nonparametric statistic when you have two independent samples.

CHI-SQUARE

This is a rather interesting statistic that deals with the probability of a certain number of occurrences in a specific category. What this statistic does is test whether or not some observed frequency of events is statistically different from what one might expect to be the frequency by chance. If you were to ask a group of scientists for their opinion about the fairness of funding for basic research you would expect there to be a fair amount of disagreement. However, the level of disagreement may be related to a specific discipline in science such as neuroscience funding versus funding for cancer or cardiovascular research. If the variables of interest are categorical, such as scientific discipline and fairness of funding, then a chi-square is the way to go. In many neuroscience paradigms categorical variables could be gender or type of drug treatment.

The key here is that the variables are categorical in nature and not quantitative. What the researcher does is count the number of times a particular event or condition occurs and tests whether or not the values observed are different from what would be expected. This statistic can be used on nominal and ordinal data.

Chi-Square Goodness-of-Fit Test

This is a use of a chi-square test to evaluate whether or not a given frequency distribution fits a predicted distribution. In a recent study it was determined that the size of synapses (based on post synaptic density size) in the frontal cortex, Brodmann area 9, has a unique distribution in older rats. The laboratory is testing a compound that can mimic a disease state that is believed to alter the distribution of synapses in the frontal cortex by eliminating some of the small and very small synapses and augmenting the number of larger contacts. The compound was given to 15 rats and the synaptic contact size was evaluated for a random sample of 100 synapses in each subject (Table 8.13). The H_0 simply states that the actual observed frequency of the distribution synapse size in the drug-treated animals is the same as that previously observed (expected) in the nontreated animals.

If we run a simple chi-square test on the data in Table 8.13 we get $X^2 = 9.92$. Checking a table of critical values for chi-square with $k - 1$ degrees of freedom (k is the total number of categories) this value is significant at the 0.05 level, indicating that the H_0 is rejected and the hypothesis that the compound alters synapse size is supported.

When reporting these results in a journal article it is important to report the actual X^2 value along with the degrees of freedom and the p values. This would look like: "A chi-square goodness-of-fit test showed that the compound had a significant effect on the synapse size ($p < 0.05$)."

TABLE 8.13   Chi-Square Goodness of Fit: Synapse Size

Synapse size	Expected percentage (%)	Observed percentage (%)
Very large	5	8
Large	20	28
Medium	40	42
Small	30	19
Very small	5	3

Chi-Square Test of Independence

The chi-square statistic can be used to test for the independence of variables. For example, you are trying to determine which therapy might enhance recovery from a moderate spinal cord injury in rats. You have an antioxidant compound that improves recovery but believe the recovery can be enhanced if combined with locomotor training or no locomotor training. To show improvement, the animals must reach a specific criterion on a mesh grid. If the animals do not reach criterion, then there is no improvement but if they do reach criterion, then there is improvement. A total of 30 animals are pretested on a mesh walking grid and subjected to a spinal cord contusion. Two days post injury they are evaluated for their locomotor ability and subsequently given the compound with or without the added locomotor training for the next 14 days. At 16 days post injury they are tested on the walking grid and compared to their scores on day 2 post trauma. Each rat is then categorized as either showing improvement or no improvement. The combined data are shown in Table 8.14.

The H_o is that there is no difference in the recovery improvement with the addition of the locomotor training. If we run a simple chi-square test on these data we get $X^2 = \mathbf{3.33}$. In testing independence in any contingency table the degrees of freedom are given by (rows -1) (columns -1). Checking a table of critical values for chi-square with 1 degree of freedom we find that a value of **3.84** is necessary for significance at the 0.05 level. Our obtained value is close to reaching this value but does not. Consequently the H_o cannot be rejected. What this essentially says is that locomotor training and recovery improvement are independent and do not have a relationship even though you might think it does. Perhaps if there were a larger number of subjects per group it would be significant. The closer the numbers in each cell are to each other, the greater the probability that they are independent. It is important to note that the degrees of freedom for the chi-square statistic differ depending on its use as a test of independence and goodness of fit. What is interesting to note is the fact that the chi-square distribution is very asymmetrical especially when the degrees of freedom are very small. As the degrees of freedom become larger the distribution becomes more symmetrical.

TABLE 8.14   Chi-Square Test of Independence: Locomotor Training

	Improvement	No improvement	Totals
Training	10	5	15
No training	5	10	15
Totals	**15**	**15**	**30**

In the above example (Table 8.14) a 2 × 2 contingency table was used. This statistic will also work for 3 × 3 or 2 × 5 contingency tables. It is not necessary that there be the same number of subjects in each group. For example, one might have 20 subjects in the training group and only 13 in the no training group. One just does not want the differences in sample size to be extreme such as 20 in one group and only 5 in another.

When reporting these results in a journal article it is important to report the actual X^2 value along with the degrees of freedom and the p values. This would look like: "A 2 × 2 chi-square test revealed that the relationship between training and locomotor activity was not significant ($p > 0.05$)."

FISHER'S EXACT TEST

This statistic is like the chi-square test, but is very useful when the number of frequencies per cell is small (i.e., one of the cells has an expected count of less than 5) and it is a 2 × 2 contingency table. Like the chi-square test, the assumption is that there is no relationship between the two variables being tested. This statistic calculates all possible outcomes by rearrangements of the observations and comparing the number of unusual rearrangements to the observed counts under the assumption that there is no association between the two variables. This statistic calculates what is known as a phi (Φ) coefficient, which is some indication of the effect size. For example, if in a set of animal experiments there is an unexpected level of death in one of the experimental conditions, and you want to determine if it may be related to one of the experimental conditions, then Fisher's exact test is appropriate. The H_o would be that there is no difference in the proportion of deaths in the two groups. You cannot use the chi-square because the expected frequencies are less than 5.

When reporting these results in a journal article, it is important to show the frequency table. It is also important to report the phi (Φ) coefficient (e.g., $\Phi = 0.648$) along with the X^2 value and the p value.

KRUSKAL–WALLIS ONE-WAY ANALYSIS OF VARIANCE

If there are only two independent groups being tested and the data probably do not follow a normal distribution then the Mann–Whitney U test (Wilcoxon rank sum test) is usually the most appropriate. However, if there are multiple groups being compared from possibly different populations, then the Kruskal–Wallis (KW) test would be the statistic of choice. This is equivalent to the parametric one-way analysis of variance.

Essentially what this test does is determine if a set of independent samples are from simply random samples from the same population or from different populations. The H_0 states that there is no difference between the groups, while the H_A states that at least one of the groups will be different from the others. Because this test is nonparametric it makes no prediction about the different population means but rather compares the medians. The sample size does not have to be the same for each group but should not differ greatly. However, this test works for small sample sizes such as 5 or 6.

During a lunch break, one of your colleagues mentions that she has discovered that several different natural compounds improve learning and believes it is due to an increase in neuronal firing rate in the CA1 region of the hippocampus. You decide to test whether or not any of these compounds can increase the firing rate and if they are different from each other. You both collaborate and collect the following neurophysiology data showing percent change in baseline firing rate using a slice preparation (Table 8.15). Different animals are used to test each compound and the values indicate the mean percent increase in firing rate for each animal. In other words, each group of compounds was tested on five different animals with a total n = 25.

Because there is wide variance in the scores, you do not think they follow a normal distribution and decide to apply the KW statistic. With the assistance of your statistical software program you determine that the KW value = 10.3 (Table 8.16). The degrees of freedom for the KW are calculated as $k - 1$, with k being the number of groups. To determine the level of significance you check the chi-square distribution table. For df = 4, a value of 9.49 is necessary to reach significance at the 0.05 level. The H_0 is not supported and you feel that the H_A is.

All this test has told you so far is that there is a difference somewhere between the different groups but not which ones (Table 8.17). To determine this, a subsequent series of tests need to be applied to the data.

There is little agreement as to which subsequent test to use and many statisticians recommend either the Dunn's test or the Mann–Whitney U

TABLE 8.15   Kruskal–Wallis ANOVA: Hippocampal Firing

KrO	SJW	CC	PC	HCE
9	8	50	29	35
5	42	45	10	48
14	33	51	35	53
13	20	15	17	26
17	30	38	19	12

TABLE 8.16 Results: Kruskal—Wallis ANOVA—Hippocampal Firing

df	4
No. groups	5
No. ties	2
H value	10.3
p value	0.0367

TABLE 8.17 Results: Kruskal—Wallis ANOVA—Hippocampal Firing

Group	n	Mean rank
KrO	5	5.3
SJW	5	13.0
CC	5	19.0
PC	5	11.2
HCE	5	16.5

statistic. Many statistical packages have these options following the KW analysis. The Dunn's test (Chapter 6) shows that only KrO and CC differ significantly. This is a rather conservative test but recommended by many statisticians. If you decide to use the Mann—Whitney U statistic then you simply have to pair each of the groups, rank the data from smallest to highest, and calculate the U score. This is repeated for each pair of scores (e.g., KrO vs SJW; KrO vs CC; PC vs HCE) and one then draws conclusions based on the outcomes (Table 8.18). As one might suspect, this can be very time-consuming if there are many groups used in the KW analysis. An alternative multiple comparison method is given by Siegel and Castellan.[1] Your friendly biostatistician can help you with this.

Often these results are graphed as shown in Figure 8.3. Since the analysis used nonparametric techniques one must plot the median and not the mean such as in this scatterplot.

When reporting the results of the KW test one must include the KW or H score, the degrees of freedom, and the p value. In the journal the text for the above experiment would be: The KW test indicated a significant effect (KW = 10.23, df = 4, $p < 0.05$) or ($H_{(4)} = 10.23$, $p < 0.05$). In the methods section it is important to state what statistic was used to contrast the different groups, i.e., Mann—Whitney U test.

TABLE 8.18   Multiple Comparisons Following Kruskal–Wallis ANOVA

Contrast	Dunn's	Mann–Whitney U
KrO-SJW	ns	ns
KrO-CC	$p < 0.05$	$p < 0.02$
KrO-PC	ns	ns
KrO-HCE	ns	$p < 0.05$
SJW-CC	ns	ns
SJW-PC	ns	ns
SWJ-HCE	ns	ns
CC-PC	ns	ns
CC-HCE	ns	ns
PC-HCE	ns	ns

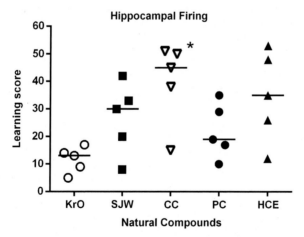

FIGURE 8.3   Note the differences in variance between the groups. Because of this hetero-geneity a nonparametric ANOVA would provide a very false analysis of the data. Each sym-bol represents a single subject's learning score. The line represents the group median. *$p < 0.05$ compared to KrO.

FRIEDMAN ONE-WAY REPEATED MEASURE ANALYSIS OF VARIANCE BY RANKS

This nonparametric test is used to compare three or more **matched** groups. It is sometimes simply called the Friedman test and often cited as Friedman's two-way ANOVA, although it is really a one-way ANOVA. **There is not a true nonparametric two-way ANOVA.** This Friedman's test is an ideal statistic to use for a repeated measures type of experiment to determine if a particular factor has an effect. As an example look at the swim speed data again in a Morris water maze (Table 8.19).

Here is a group of six different rats with their swim speeds on different days as a result of a change in the water temperature. The question is whether or not changing the water temperature in the maze affects the swimming speed. The H_o is that the swim speed will be the same regardless of the water temperature. Essentially what this test does is rank the swim speeds for each subject across the various water temperatures and in this way each animal is compared to itself. The magnitude of the differences with each subject is not important. This test results in a Friedman statistic (F_r) and the possible significance of this value is then looked up in a specific table literally called the critical values for the Friedman two-way analysis of variance by ranks statistic, which is available on the web. In this experiment there are four treatments with a total of six subjects and the $F_r = 14.90$. In this table a value of 7.6 is necessary for $\alpha < 0.05$ and 10.0 for significance at the 0.01 level. Thus the H_A is supported. A Dunn's test is used to subsequently compare the different treatments following the Friedman analysis. In this particular example, the 20 °C temperature was significantly different from the 27 and 32 °C temperatures but not the 24 °C.

When reporting the results of the Friedman test one must include the number of subjects, the F_r score, and the p value. Some journals suggest that the degree of freedom (here 4,6) should also be given. This might

TABLE 8.19   Friedman Repeated Measure ANOVA: Swim Speed

Animal	20 °C	24 °C	27 °C	32 °C
1	39	38	34	33
2	29	25	20	20
3	36	37	29	24
4	25	29	18	19
5	31	27	24	22
6	34	33	30	31

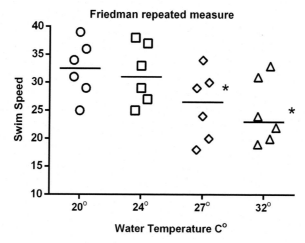

FIGURE 8.4   The nonparametric version of a repeated measures one-way ANOVA is the Friedman test. Each symbol represents an individual data point for each subject under different water temperature conditions. The horizontal line represents the group median. *$p < 0.05$ compared to the 20 °C cohort.

appear as: "a Friedman analysis of variance was applied and indicated that a change in water temperature significantly altered the swim speeds ($F_r = 14.90$, df 4,6, $p < 0.01$). A Dunn's test revealed that the swim speed in 20 °C temperature water was significantly increased compared to the 27 and 32 °C water temperatures." These data can be graphed as shown in Figure 8.4 using a scatterplot and the median.

SPEARMAN'S RANK ORDER CORRELATION

This nonparametric statistic, also known as Spearman's rho, is the equivalent of the parametric Pearson correlation coefficient. The Spearman's rho is named after Charles Spearman, an experimental psychologist at the University College of London, who invented the procedure in 1904. It is used when one or both of the variables are ordinal scaling. It is based on ranks of the data and not on the data itself and thus is resistant to outliers. This is a very important feature since outliers can significantly alter the outcome of the Pearson statistic (Chapter 5). The H_0 is that the two variables are independent and have no relationship to each other. Like the Pearson statistic, the range of the correlation is +1 to −1 with a zero indicating no correlation between the two variables. An added characteristic of the Spearman's rho is that it does not assume that the correlation will be linear. Consequently, one can use this statistic even if the relationship is curved. The one basic assumption is that the underlying relationship is **monotonic**, which means that as one variable becomes larger the other becomes consistently either larger or smaller. Statisticians might say that the two variables under investigation covary, which means simply that as one variable increases, the other variables will either increase or decrease. A **nonmonotonic** relationship would be one where one variable becomes larger while the other sometimes becomes larger and then sometimes becomes smaller. The other basic assumption for using the Spearman is that the observations are independent. The interpretation of Spearman's rho is the same as that for the Pearson statistic. Let us take the example of examining the possible association between synaptic numbers in a small part of the frontal cortex and an individual's global cognitive score (Table 8.20).

This particular set of scores includes both ordinal and ratio types of data. There is also a considerable range in the scores, making the use of

TABLE 8.20   Spearman's Rho Correlation: Synapses × Cognition

Subject	Synapses	Global cognition
1	380.1	26
2	415.2	24
3	381.1	16
4	548.5	27
5	700.5	26
6	354.1	13
7	418.5	27
8	525.0	25
9	532.3	23
10	497.3	21
11	169.8	9
12	306.5	24
13	378.6	28
14	121.6	11
15	296.8	15
16	431.7	28
17	387.2	23
18	494.6	20
19	433.2	16

Spearman's rank correlation ideal for these data. This statistic can also be used with interval scale data. What this test does is first rank all of the data from highest to lowest in terms of one of the variables. Most statisticians use the ordinal variable to rank the data (Table 8.21). The highest score is assigned a 1 and then a 2 to the next highest score and so on. When there are tied scores involved, these are all assigned the same average rank. For example, in this set of data, the first and second scores are a 28 so they both get 1.5. This rank of 1.5 would occupy both the first and the second places in the list of scores. It is important to note that if there are too many tied scores it will affect the size of rho (ρ), the end statistic.

Almost every statistical software program has the option of using this nonparametric statistic. The printout may differ but should contain information such as the "r" or rho value and the p value, usually a

TABLE 8.21   Spearman's Rho Correlation: Synapses × Cognition

Rank	Global cognition	Synapses
1.5	28	431.7
1.5	28	378.6
3.5	27	548.5
3.5	27	418.5
5.5	26	700.5
5.5	26	380.1
7	25	525.0
8.5	24	415.2
8.5	24	306.5
10.5	23	532.3
10.5	23	387.2
12	21	497.3
13	20	494.6
14.5	16	381.1
14.5	16	433.2
16	15	296.8
17	13	354.1
18	11	121.6
19	9	169.8

TABLE 8.22   Results: Spearman's Rho
Correlation—Synapses × Cognition

Spearman's r	0.475
p value	0.05
Rho corrected for ties	0.473
Tied p value	0.05
No. of ties global cognition	6

two-tailed p value (Table 8.22). If there are ties, it is important to note the tied r value and the tied p value. For this example, what we can say is that the relationship between the number of synapses in the frontal cortex and the performance on a global cognition test for the 19 subjects is 0.473, which is significant at $p < 0.05$. If the H_o stated that there was no relationship then these results would support the H_A.

When reporting the results of the Spearman's rho, it can be graphed in the same way as the Pearson correlation (Figure 8.5). In the text it is important to include the number of independent measures (here the n = 19), the names of the two variables being evaluated (global cognition; synapses), the value of rho ($r_s = 0.475$), and the p value ($p < 0.05$). Some journals also require stating the number of ties (6) in the analysis and the

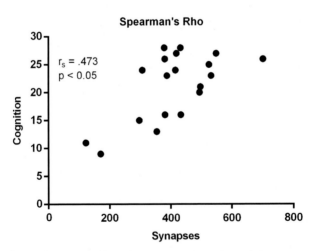

FIGURE 8.5   Spearman's rho (r_s) statistic is the equivalent of the parametric Person's product—moment correlation. Like the parametric correlation, it is improper to place a regression line on the graph. Each circle represents an individual data point.

degrees of freedom. For the Spearman's rho, the degrees of freedom are simply the number of pairs in the sample minus 2 (n − 2). For this example df = 17. Reports in the literature might include: "There was a significant correlation between the individual's global cognition score and the synaptic counts in the superior frontal cortex [r_s (17) = 0.475, $p < 0.05$]. There were six tied scores in the data."

One final word on reporting the results of the Spearman's rank order correlation—when graphing the results, it is inappropriate to place a regression line on a graph because this line indicates a cause and effect relationship, which may or may not be the case. This is the same as that discussed in Chapter 5 for the Pearson rho. Some authors place a dotted "regression" line and explain in the figure legend that this line merely indicates the direction of the association and does not imply cause and effect. Some journals allow this and others do not.

KENDALL RANK ORDER CORRELATION COEFFICIENT

This nonparametric statistic is very similar to Spearman's rho. It is sometimes referred to as Kendall's tau (T). It is calculated differently from Spearman's rho and consequently has different underlying scales resulting in different correlation values. If the data concerning global cognition and synaptic numbers are analyzed with Kendall's T one would get the results shown in Table 8.23.

The interpretation of Kendall's T is complicated and different from that of Spearman's rho. Spearman's rho can be interpreted the same as the parametric Pearson product—moment correlation in terms of the proportion of the variability accounted for by each variable, as described above. Kendall's T deals with the probability that the two variables **are** in a specific order as opposed to the probability that they **could** be in a different order. A more detailed explanation is beyond the scope of the present discussion. This statistic is not used very often and researchers

TABLE 8.23   Kendall Rank Order Correlation Coefficient

Kendall's T	0.368
p value	0.028
Rho corrected for ties	0.375
Tied p value	0.025
No. of ties global cognition	6

who wish to apply it to their data should first consult with a biostatistician. It is inappropriate to analyze some of the data with Kendall's T and some with Spearman's rho.

NONPARAMETRIC AND DISTRIBUTION-FREE ARE NOT REALLY THE SAME

The actual terms nonparametric and distribution-free are not completely synonymous but the popular statistical press has reinforced this idea. One must remember that the nonparametric tests, just like their counterparts such as the t-test and the F-test, are really only estimations of the population as characterized by the sample. When interval or ratio data are converted to ranks or frequencies and evaluated by nonparametric techniques they lose power. There is a greater chance of a Type II error. One also must remember that some of the information used when collecting these data has also been lost in the sense that it no longer plays the same influential role in the interpretation of the results.

SUMMARY

- Nonparametric statistics should be used when the data contain a large amount of variance in one or more groups.
- Nonparametric tests can be used on very small samples.
- For every parametric statistic, with the exception of the two-way ANOVA, there is an equivalent nonparametric test.
- One should never report the mean and standard deviation when using a nonparametric statistic.

References

1. Siegel S, Castellan NJ. *Nonparametric Statistics for the Behavioral Sciences*. Boston: McGraw Hill; 1988.
2. Hollanders M, Wolfe DA. *Nonparametric Statistical Methods*. New York: John Wiley & Sons; 1999.
3. Corder GW, Foreman DI. *Nonparametric Statistics for Non-statisticians*. New York: John Wiley & Sons; 2009.

Outliers and Missing Data

Stephen W. Scheff

University of Kentucky Sanders–Brown Center on Aging, Lexington,
KY, USA

If something has a 50% chance of happening, then 9 times out of 10 it will.
Yogi Berra

Fundamental Statistical Principles for the Neurobiologist
http://dx.doi.org/10.1016/B978-0-12-804753-8.00009-9

183

In "straight" statistical terms, an **outlier** is an observation or data point that "appears" to deviate markedly from the other observations in the data set, and may be a data point that belongs to a totally different population than the one being examined. It can either be too large or too small. Methods for dealing with outliers are again one of the heavily debated issues in statistics and there are several books on this topic such as the one by Barnett and Lewis.[1] Another term for this is called cleaning up your data set so it makes biological sense. Almost every neuroscientist has seen data sets with outliers and really not sure how to deal with them. Let me state quite **BOLDLY** and with **EXTREME** emphasis, it is _**NOT acceptable**_ to simply remove a data point because you think it is an outlier or because it "ruins" your hypothesis. Why is it important to identify even a single outlier? One of the most common examples is how a single outlier can totally change a correlation coefficient. The Pearson product–moment correlation is one of the most widely used statistics that measures the linear relationship between two variables. It is often used to evaluate not only the reliability of data but often its validity (Chapter 5). However, a single outlier in a data set can dramatically alter the outcome of this statistical analysis as shown in Figure 9.1. The single data point made a very weak relationship look very strong and significant.

There is an interesting law in the field of statistics that basically says, "If you increase the sample size you get a greater approximation of the true population." While it is true that larger numbers can greatly reduce the variance, one should never underestimate the ability of a single outlier to warp a data set.

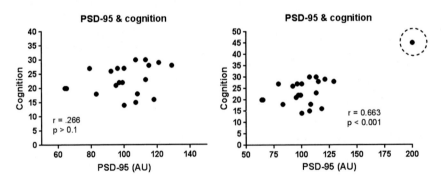

FIGURE 9.1 The correlation on the left is graphed without any outliers and this correlation shows a _p_ value greater than 0.1. A single outlier (broken circle), as seen in the graph on the right, can totally distort the correlation, which is not significant at the $p < 0.001$ level.

REASONS FOR OUTLIERS

There are multiple different reasons why one has an outlier or several outliers.

1. The data were simply entered into the data set incorrectly such as with transposed numbers: 187, 178, 189, 192, **814**, 162, 154, 185. Before getting all excited, simply carefully check each entry in the data set and verify all the numbers. Sometimes a mathematical calculation was made on each value before entering values. Go back and recheck all the calculations associated with that data point. The common term for this is **Data Quality Assessment** (DQA).
2. Unintentional differences during the experimental procedure. Often an experiment has multiple steps especially neurochemistry experiments. There could easily have been a pipetting error or an incorrect concentration. These types of errors can have devastating consequences and thus all data values must have adequate controls. If a value in the data set appears out of place, then carefully check laboratory notes and check other values that were collected at the same time to see if there are identical types of peculiarities.
3. Misidentification of subjects within a group. It is a common statistical practice to be blinded with respect to group designation during the data collection phase. A good laboratory practice is to code subject material prior to data collection and then to decode group designation only after the collection phase is finalized. It may be the case that a particular subject was initially coded incorrectly or the subject identifier at the time of experimentation was misread. Careful recheck of subject identifiers immediately prior and after the experiment can often identify a reason for an outlier.
4. Biological variance may be responsible for an aberrant data point. In this particular case the data point may truly not be representative of the population under investigation but may be very valid.

REMOVING OUTLIERS

Initially one must deal with the question of whether or not it is ok to remove outliers. One should first try to fix the mistake such as a simple transposed number or perhaps a missing value as discussed above. Was the sample possibly contaminated before or during the data collection phase and if so can it be rerun? Although some scientists simply believe that eliminating outliers is deceitful, one can also argue that not removing them is also portraying fraudulent conclusions. I am going to take the stance that true outliers should be eliminated from a data set but still noted in the laboratory notebook. Let me make one point very clear: **It is a mistake to run an analysis and then subsequently delete data points because you now consider them to be outliers**. Ad hoc removal of data points, especially if their inclusion goes against your hypothesis, is unacceptable.

Peirce's Criterion for Data Elimination

This particular practice was established by Benjamin Peirce in 1852.[2] He taught at Harvard for over 50 years and was both a mathematician and an astronomer (which was common at the time). His method, which is based on probability theory, is really directed at determining if a data point truly is an outlier. Here is a quote from Peirce's article that appeared in *The Anatomical Journal* in 1852 entitled, *Criterion for the rejection of doubtful observations.*

"The principle upon which it is proposed to solve this problem is, that the proposed observations should be rejected when the probability of the system of errors obtained by retaining them is less than that of the system of errors obtained by their rejection multiplied by the probability of making so many, and no more, abnormal observations" (Astronomical Journal, 1852[2]).

Here is how Peirce's criterion for data elimination works for a set of values. We suspect that the value 162 may be an outlier in Table 9.1.

First determine the mean and standard deviation of the set of numbers:

$$\overline{X} = 222.5, \; SD = 30.8$$

TABLE 9.1   Data with One Possible Outlier

Subject	Score
1	193
2	212
3	264
4	243
5	251
6	162
7	210
8	209
9	237
10	244

Go to the table for critical values of Peirce's R Table 9.2 and find the value assuming 1 measured outlier quantity for 10 measurements since there are a total of 10 numbers being evaluated.

$$R = 1.878$$

Calculate the maximum allowable deviation $|x_i - x_m|\text{max} = \text{SD}\cdot R = 30.8(1.878) = 58.03$.

This number, 58.03, is the "magic" value for data elimination. If the difference between the mean of the values and the suspected outlier is greater than this number, it should be eliminated.

$$|x_i - x_m| = \text{Dev}_{\text{act}}$$
$$|162 - 222.5| = 60.5$$

TABLE 9.2   Peirce's Table for R

N	1	2	3	4	5
3	1.216				
4	1.383	1.078			
5	1.509	1.200			
6	1.610	1.299	1.099		
7	1.693	1.382	1.187	1.022	
8	1.763	1.453	1.261	1.109	
9	1.824	1.515	1.324	1.178	1.045
10	1.878	1.570	1.380	1.237	1.114
11	1.925	1.619	1.430	1.289	1.172
12	1.969	1.663	1.475	1.336	1.221
13	2.007	1.704	1.516	1.379	1.266
14	2.043	1.741	1.554	1.417	1.307
15	2.076	1.775	1.589	1.453	1.344
16	2.106	1.807	1.622	1.486	1.378
17	2.134	1.836	1.652	1.517	1.409
18	2.161	1.864	1.680	1.546	1.438
19	2.185	1.890	1.707	1.573	1.466
20	2.209	1.914	1.732	1.599	1.492
21	2.230	1.938	1.756	1.623	1.517

Continued

TABLE 9.2   Peirce's Table for R—cont'd

N	1	2	3	4	5
22	2.251	1.960	1.779	1.646	1.540
23	2.271	1.981	1.800	1.668	1.563
24	2.290	2.000	1.821	1.689	1.584
25	2.307	2.019	1.840	1.709	1.604
26	2.324	2.037	1.859	1.728	1.624
27	2.341	2.055	1.877	1.746	1.642
28	2.356	2.071	1.894	1.764	1.660
29	2.371	2.088	1.911	1.781	1.677
30	2.385	2.103	1.927	1.797	1.694
40	2.504	2.230	2.059	1.932	1.832
50	2.592	2.326	2.158	2.035	1.936
60	2.663	2.401	2.237	2.116	2.019

This partial table was modified from the one published by Gould.[11]

Since 60.5 > 58.03 the data point should be eliminated and considered an outlier.

Let us suppose that there are two data points that "appear" suspicious such as in Table 9.3.

The process is very similar to that used for just a single possible outlier with the exception that the R value from Peirce's table now becomes **1.570** instead of 1.878 (Table 9.4).

Determine the mean and standard deviation of the set of numbers: $\overline{X} = 232.5$, SD $= 36.07$.

Calculate the maximum allowable deviation $|x_i - x_m|$max $=$ SD·R $= 36.07(1.570) = 56.63$.

Actual deviations for "suspicious" data:

$$|x_i - x_m| = \text{Dev}_{act}$$
$$|293 - 232.5| = 60.5$$
$$|162 - 232.5| = 70.5$$

Since 60.5 > 56.63 and 70.5 > 56.63, both data points should be eliminated and considered outliers. The new mean and standard deviation for the set of eight numbers are $\overline{X} = 233.75$, SD $= 20.9$.

If the Peirce method is used, it must be stated very clearly in the methods section how many data points were eliminated.

TABLE 9.3   Data with Two Possible Outliers

Subject	Score
1	293
2	212
3	264
4	243
5	251
6	162
7	210
8	209
9	237
10	244

TABLE 9.4   Peirce's Table for R

N	1	2	3	4	5
3	1.216				
4	1.383	1.078			
5	1.509	1.200			
6	1.610	1.299	1.099		
7	1.693	1.382	1.187	1.022	
8	1.763	1.453	1.261	1.109	
9	1.824	1.515	1.324	1.178	1.045
10	1.878	1.570	1.380	1.237	1.114
11	1.925	1.619	1.430	1.289	1.172
12	1.969	1.663	1.475	1.336	1.221
13	2.007	1.704	1.516	1.379	1.266
14	2.043	1.741	1.554	1.417	1.307
15	2.076	1.775	1.589	1.453	1.344

This partial table was modified from the one published by Gould.[11]

Chauvenet Criterion

This particular method for identifying outliers was first described by William Chauvenet in 1863.[3] Like Peirce, he was a mathematician and astronomer. This method is fairly simple but one needs a special table in which to calculate what is called the tau$_{max}$ (τ_{max}).

Here is how the Chauvenet criterion works for data elimination for the same set of values used above. We suspect that the value 162 in Table 9.5 may be an outlier.

First determine the mean and standard deviation of the set of numbers: $\overline{X} = 222.5$, SD $= 30.8$. Since there are 10 data points, one needs to find the "Ratio of Maximum Acceptable Deviation to Precision Index (d_{max}/S_x)" from the Chauvenet table of t_{max}.

To calculate the ratio simply use the equation $P_z = 1/(4n)$.

P_z is the probability represented by one tail in a normal distribution and n is the sample size.

$$P_z = 1/(4n) = 1/40 = 0.025$$

Next simply look up the value beyond the area under the standard normal distribution for 0.025, which is a Z score of 1.96.

A portion of that table is shown in Table 9.6.

The SD(t_{max}) $= 30.8 \ (1.96) = 60.4$.

The allowable difference between the data point in question and the mean must be <60.4 to be retained.

$$|162 - 222.5| = 60.5$$

Since $60.5 > 60.4$ the data point is rejected.

TABLE 9.5   Data with One Possible Outlier

Subject	Score
1	193
2	212
3	264
4	243
5	251
6	162
7	210
8	209
9	237
10	244

TABLE 9.6   Chauvenet Criterion for Rejecting a Data Point

Number of data points	Ratio of maximum acceptable deviation to standard deviation, d_{max}/SD
3	1.38
4	1.54
5	1.65
6	1.73
7	1.81
8	1.86
9	1.91
10	1.96
11	2.00
12	2.04
13	2.07
14	2.10
15	2.13
16	2.15
17	2.18
18	2.20
19	2.22
20	2.24
25	2.33
30	2.39
40	2.49
50	2.57
100	2.81

If a particular data set has an n that is not listed in the above table, one simply has to interpolate the difference between the two closest numbers shown in the table (e.g., 23 data points = a t_{max} of 2.29). There are a number of examples of the Chauvenet rejection table on the web.

There should be some real caution about removing outliers when the total number of values considered is less than 10. The problem is that the standard deviation estimated from a small sample is also markedly influenced by the outlier and may subsequently provide an unrealistic estimate of the population variance. One might consider using a nonparametric statistic.

The Chauvenet method has come under considerable criticism because it makes an arbitrary assumption about the level of deviation for the rejection of a data point. The chance of wrongly rejecting a single data point could be as high as 40%. So proceed with caution with this technique.

Grubbs' Test for Outliers

The Grubbs' test statistic was described by Frank Grubbs[4,5] and is also known as the **Extreme Studentized Deviate test (ESD)** or **Maximum Normal Residual test (MNR)**. This test is based on the assumption of normality. Many statistical packages have a routine for this procedure. Basically one calculates a Z value, which is the difference between the "suspicious" data point and the sample mean divided by the standard deviation of all the scores including the data value in question. Using the same example from above (Table 9.1) for the data value of 162, the Z value $= 1.96$.

$$Z = \frac{|\text{Mean} - \text{Suspicious data point}|}{\text{SD}} = \frac{|222.5 - 162|}{30.8} = \mathbf{1.96}$$

Consulting Table 9.7 for critical values for a sample size of 10, the calculated Z value needs to equal or exceed **2.176**. Since it does not, the data point is **not considered an outlier**.

There are a number of web sites that have a calculator for the Grubbs' test.

Dixon Q Test

This is a relatively common outlier test, developed by Wilfrid J. Dixon,[6] that is very simple to run but considered to be quite conservative and can only be used for a data sample size 30 or less and some say less than 10. It is sometimes called simply the **Dixon test** or **Q test**. It assumes a normal distribution and should be used no more than once on a data set. To run this test one simply arranges the values from highest to lowest and uses the following to determine the Q value:

$$Q = \frac{|\text{outlier} - \text{closest value}|}{|\text{range}|}$$

TABLE 9.7 Critical Values for the Extreme
Studentized Deviate (Grubbs') Test[a]

N	0.05	0.01
5	1.672	1.749
6	1.822	1944
7	1.938	2.097
8	2.032	2.221
9	2.110	2.323
10	2.176	2.410
11	2.234	2.485
12	2.285	2.550
13	2.331	2.607
14	2.371	2.659
15	2.409	2.705
16	2.443	2747
17	2.475	2.785
18	2.504	2.821
19	2.532	2.854
20	2.557	2.884
21	2.580	2.912
22	2.603	2.939
23	2.624	2.963
24	2.644	2.987
25	2.663	3.009
26	2.681	3.029
27	2.698	3.049
28	2.714	3.068
29	2.730	3.085
30	2.745	3.103
35	2.811	3.178
40	2.866	3.240
45	2.914	3.292

Continued

TABLE 9.7   Critical Values for the Extreme
             Studentized Deviate (Grubbs')
             Test[a]—cont'd

N	0.05	0.01
50	2.956	3.336
55	2.992	3.376
60	3.025	3.411
65	3.055	3.442
70	3.082	3.471
75	3.107	3.496
80	3.130	3.521
85	3.151	3.543
90	3.171	3.563
95	3.189	3.582
100	3.207	3.600
110	3.239	3.632
120	3.267	3.662

[a]*These values are for a two-tailed test.*
*This partial table was modified from the one published by Grubbs and
Beck.[12]*

Using the example above (Table 9.1) for value **162** in the data set of 10
values

$$162, 193, 209, 210, 212, 237, 243, 244, 251, 264$$

$$Q = \frac{|162 - 193|}{|264 - 162|} = \frac{31}{102} = 0.304$$

Consulting a table for critical values for the Dixon Q test, which can be
found on the web, for a sample size of 10, we find that a value of 0.466 is
necessary to be 95% confident that the value is an outlier and a value of
0.568 at the 99% confidence level. Since our computed Q value is 0.304, it
is not an outlier. In fact, even at the 90% confidence level it is not
considered an outlier since the Q value is 0.412.

The same procedure can be used if the suspected outlier is the highest
value in the data set. For example, if we had the following values

$$193, 209, 210, 212, 237, 243, 244, 251, 264, \textbf{334}$$

$$Q = \frac{|264 - 334|}{|193 - 334|} = \frac{70}{141} = 0.496$$

In this case we would say with 95% confidence that the value **334** was an outlier but could not be 99% certain since the critical value would be equal or exceed 0.597. There is an interesting paper by Rorabacher[7] that reported finding some errors in the original paper by Dixon and subsequently published a correct set of tables for this statistic.

Eliminating Values Greater than Two Standard Deviations

Why cannot one just look at the data set and decide that any point that is more than 2 standard deviations from the mean should be removed? In a normal Gaussian distribution, 95% of the data points should be within 2 standard deviations and the probability of a single value not being within that range is 5%. That means that 1 in 20 measurements should be outside the 95% interval and thus part of the normal Gaussian distribution and should not be considered an outlier. With even a smaller set of data points the probability is increased and your chances of discarding an important data point are much greater. In the example in Table 9.1, 2 standard deviations would make a range of 160.9−284.1. The value of 162 would fall within the range and not be considered an outlier. However, if the data set included the value of 152, the mean would be 221.5 with a standard deviation of 33.18. The Grubbs' test would not consider the data point as an outlier even though the Z value was 2.09. However, the 2 standard deviation range would be 155.14 to 287.86 and one would reject the value of 152 since it would fall outside this range.

One must decide for themselves if a value is considered an outlier before data analysis is run. There is no rule as to which method one chooses. Many scientists simply like to be very conservative and include as many data points as possible, in which case the Grubbs' test would be the choice. After running the Grubbs' test one is not allowed to now search for another outlier because the standard deviation has changed significantly. Whatever method is chosen it must be clearly stated in the methods section.

Here is a word of caution when removing data points. Most scientists design experiments with equal numbers of subjects in each group. Most statistics are designed for an equal number of subjects per group (balanced design). When removing a data point you sometimes invalidate some statistical procedures. Whenever I read a study where the sample sizes are different between groups, and nothing is noted in the results section about the loss or removal of subjects, I get very suspicious about those data. Ask yourself why someone would design an experiment with an n = 7 in one group and an n = 9 in another. Sometimes the differences are even larger such as an n = 4 in one group and n = 8 in another group.

MISSING DATA

Invariably, in the course of running experiments, there is a lost data point or points. The reasons for this could be, because there was a malfunction in a piece of equipment, a tissue section on a slide was not stained properly, an animal was not able to perform on a given day, a technician forgot to enter a value in a spreadsheet, etc. It is not uncommon in the collection of human subject information that a respondent will not answer a specific question or will provide an answer that does not fit into one of several predetermined categories. An example of this is identifying the race of an individual. Usually there are categories such as white (non-Hispanic), white (Hispanic), African American, Asian, American Indian. What happens if someone enters "human" or simply leaves the space blank? Should this eliminate all of the data from the subject? The bottom line is that it happens and as a neuroscience researcher you have to deal with it.

The major problem in statistics, when missing data occur, is that most of the techniques for analyzing data require equal numbers in a complete data set. Most statistical packages simply deal with a missing data point by eliminating that particular subject from the analysis. This could have important ramifications if there are limited numbers of subjects in a complex set of experiments. These statistical programs simply assume complete data sets. A loss of a subject means a loss of statistical power for that analysis.

Listwise Deletion

The traditional method often used by statistical software is called **listwise deletion**. If subject No. 14 in the data set is missing an important data point, that subject is simply removed from the total analysis. The end result of such an analysis is that the group mean will be different as well as the standard deviation. If more than a single data point is missing in a given data set, then the loss of power could be substantial, depending on the overall number of subjects. If the overall size of the sample is very large and the number of missing data points is very small relative to the overall sample, then subject deletion may not be a major contribution to the overall interpretation of the data. However, consider the possibility of the loss of two subjects when there are only 10/group. A loss of two subjects in a group would be a loss of 20%, which is a significant change in the overall sample mean and variance.

Pairwise Deletion

This method actually only applies to a specific type of data set where multiple different dependent variables are used from a single subject. For example, the verbal leaning scores are being evaluated in a large group of individuals who have had mild, moderate, or severe head trauma. One of the variables of interest is the level of education of the subjects, along with time postinjury, and type of therapy postinjury. For a couple of the subjects the level of education is missing. In a pairwise deletion technique, instead of eliminating those subjects who do not have a complete data set (listwise deletion), they are only removed in the analysis that involves the use of education and still retained in those statistics that do not use that variable. The problem with this method is that some of the results will be based on different sample sizes and subsequently different standard errors. It would be very important to state in the results section that there were missing data points. This becomes very important in correlation studies.

Mean Substitution

In this scenario one simply replaces the lost data point with the mean of the remaining data. The theory behind doing this is that the group mean is the best estimate of a particular subjects score. The major problem with this is the creation of a more inaccurate estimate of the population. The addition of the group mean as a replacement results in absolutely no new information with the added feature that it reduces the variance in the sample. If you take a group of numbers (12, 19, 14, 22, 10, 16) and calculate the mean (15.5) and then add that number of the group and calculate the mean, you get exactly the same number (15.5). All it does is reduce the estimate of the error. Thus, this is not an appropriate way to deal with a missing value.

Linear Regression Substitution

In this approach, the existing data are used to plot a regression of the independent variable and the dependent variable and uses the regression equation to predict the missing data point. The theory behind this is that by knowing how the existing variables relate will provide the most accurate information for the missing data point. This method is better than the mean substitution because it is at least conditional on the other available information. However, it suffers from the same problem as the mean substitution in that it does not really add any new information. What it does is add information that is in good agreement with the existing scores. This will have an effect of reducing the estimate of the error.

Identical Characteristic Substitution

In this procedure, various subject characteristics are evaluated and the score of another subject with those same or very similar characteristics is substituted for the missing value. For example, an experimenter has collected at autopsy samples of lamina II of several different brain individuals with Pick's disease. He is interested in comparing the synaptic numbers in these subjects to age- and postmortem-matched individuals with no dementia. However, at the time of analysis he realizes that he does not have a value for one of the Pick's disease subjects for Brodmann's area 37. He is unable to go back and reanalyze that tissue. Instead, he looks over all of the dementia cases and finds another individual that is the same age and has the same postmortem interval as the subject he is missing data from so he uses that same score to fill in the missing data point. This is an inappropriate substitution because it makes the assumption that those characteristics (age and postmortem interval) are accurate predictors of the missing data. In essence, it is a very biased method. It also makes an assumption that these two individuals will respond to the disease in exactly the same way when in fact the duration of the disease between the two individuals may be very different.

Random Choice Substitution

In this method, the existing members of the group are assigned a number and a random number generator then picks one of the numbers. That value is substituted for the missing value in the data set. This method has an advantage over the other substitution methods in that it has the possibility of not only changing the group mean but also the estimate of the error. If a single value is missing from a data set, the random choice substitution method is the easiest and least problematic method.

Some of the sophisticated computer programs have some rather elaborate methods for dealing with missing data that are beyond the scope of this book. If the missing data are extremely important (actually one could say that all data points are important) then it would be most beneficial to approach your friendly neighborhood statistician and work with him or her. The best practice is to try and eliminate the problem in the first place. There actually is not any real substitution of missing data. If possible, use statistical methods that allow for unequal subjects per group. The overall influence of missing data will actually be dictated by the overall sample size. It is absolutely imperative that missing data and the method used to deal with it be stated clearly in the methods or results section of the manuscript. If you used one of the substitution methods (e.g., random choice), be very specific how it was done. If no data are missing, then that should be stated very clearly.

There are several excellent discussions on this topic of missing data and the inquiring student may want to check out the following books: Allison[8]; Cohen et al.[9]; Graham.[10]

Here is one final word about outliers and missing data. **It is totally unacceptable** to use a data elimination procedure first to remove an outlier, and subsequently to use a missing data procedure to replace it.

SUMMARY

- You should always try to use all of the collected data.
- While it is sometimes necessary to remove a data point, the method chosen should be the most conservative.
- When data points have been removed it should be clearly stated in the methods section which technique was used.
- Although missing data points are problematic for some statistical tests, one should first consider using a statistic that is adapted for unbalanced designs.
- Random choice substitution is one of the most conservative methods for dealing with missing data.
- **It is totally unacceptable** to use a data elimination procedure first to remove an outlier, and subsequently to use a missing data procedure to replace it.

References

1. Barnett V, Lewis T. *Outliers in Statistical Data.* Chichester, UK: John Wiley & Sons; 1994.
2. Peirce B. Criterion for the rejection of doubtful observations. *Astron J.* 1852;45:161–163.
3. Chauvenet W. *Manual of Spherical and Practical Astronomy.* In: *Theory and Use of Astonomical Instruments.* vol. II. Philadelphia: J.B. Lippincott; 1863.
4. Grubbs FE. Sample criteria for testing outlying observations. *Ann Math Stat.* 1950;21: 27–58.
5. Grubbs FE. Procedures for detecting outlying observations in samples. *Technometrics.* 1969;11:1–21.
6. Dixon WJ. Processing data for outliers. *Biometrics.* 1953;9:74–89.
7. Rorabacher DB. Statistical treatment for rejection of deviant values: critical values of Dixon's "Q" parameter and related subrange ratios at the 95% confidence level. *Anal Chem.* 1991;63:139–146.
8. Allison PD. *Missing Data.* Newbury Park, CA: Sage; 2002.
9. Cohen J, Cohen P, West SG, Aiken LS. *Applied Multiple Regression/Correlation Analysis for the Behavioral Sciences.* Mahwah, NJ: Lawrence Erlbaum Associates; 2003.
10. Graham JW. *Missing Data: Analysis and Design.* New York: Springer; 2012.
11. Gould BA. On Peirce's criterion for the rejection of doubtful observations with tables for facilitating its application. *Astron J IV.* 1855;83:81–87.
12. Grubbs FE, Beck G. Extension of sample sizes and percentage points for significance tests of outlying observations. *Technometrics.* 1972;14:847–854.

10

Statistic Extras

Stephen W. Scheff

University of Kentucky Sanders–Brown Center on Aging,
Lexington, KY, USA

Knowin' all the words in the dictionary ain't gonna help if you got nottin' to say. **Blind Lemon Jefferson**

STATISTICS SPEAK

Acceptance region statistical outcomes that lead to acceptance of the null hypothesis.

Alias an effect in a fractionally replicated design that cannot be distinguished from another effect.

Alpha error Type I error, when the null hypothesis is rejected when in fact it is true.

Alternative hypothesis the hypothesis that is the opposite of the null hypothesis and which is accepted when the null hypothesis is rejected.

Bell-shaped curve a very symmetrical frequency distribution of data points with a single peak when graphed; sometimes called a normal distribution.

Beta error Type II error, when the null hypothesis is accepted when in fact it is false.

Bias any effect that systematically distorts the outcome of experiments that makes the sample not representative of the population.

Fundamental Statistical Principles for the Neurobiologist
http://dx.doi.org/10.1016/B978-0-12-804753-8.00010-5

201

Biased sample a sample that is drawn in which each member of the population did not have an equal chance of being selected for the sample.

Bivariate involving two variables.

Causal study measures the outcome of one variable relative to a totally different variable.

Coefficient of determination a measure of the proportion of variability that is shared by two variables.

Concomitants variables that are unaffected by various treatments (e.g., age, gender).

Confidence interval a range of values that, considering all possible samples, has some designated probability of including the true population value.

Confounding a procedure whereby treatments are assigned to subjects so that certain effects cannot be distinguished from other effects.

Consistent estimator an estimating procedure is consistent if the estimates it yields tend to approach the parameter more and more closely as the sample size approaches infinity.

Contrast comparison of two groups.

Critical region a set of outcomes of a statistical test that lead to the rejection of the null hypothesis.

Critical value the value of a statistic that corresponds to a given significance level determined from a sampling distribution.

Cumulative relative frequency the sum of all the frequencies at or below a given value represented as a percentage of the total number of data points; example: the 47 male rats in the first room and 93 male rats in the second room of the vivarium have a cumulative relative frequency of 0.27; from this we can estimate that the total number of male rats in the vivarium is about 526.

Degrees of freedom number of independent observations in the calculation of a statistic that are free to vary without affecting the total.

Dependent variable the parameter that you are interested in measuring as a result of manipulation of the independent variable; example: following treatment with a drug or vehicle, the number of correct choices a subject makes in a maze.

Directional test one-tailed test; prediction that one value is higher than another value.

Distribution-free method a method for testing a hypothesis that does not rely on the form of a specific underlying distribution; nonparametric statistics are often described as distribution-free statistics.

Expected value the average value of a random variable that would be obtained from an indefinite number of samplings of a specific population.

External validity the extent that a study can be generalized to other groups or investigations.

Extraneous variables variables that the investigator does not want to include in the design of the study, and wants to control because they could possibly contaminate the interpretation of the study.

Familywise error rate the probability that at least one of a group of comparisons will include a Type I error.

Frequency distribution a classification scheme of placing raw data into predefined groups based on some unique characteristic; example: the number of male rats in different rooms in the vivarium. Room 1 (47); 2 (93); 3 (82); 4 (113); 5 (147); 6 (44).

Fixed effects model an experimental design in which all treatment levels of a specific variable are deliberately arranged by the experimenter.

Histogram graphical display of a frequency distribution.

Homogeneity of variance refers to the amount of spread of the data between samples, and it means that the samples have similar amounts of variance.

Incomplete block design a block experimental design that does not include all treatment levels or combinations of treatment levels contained in an experiment.

Independent variable the main parameter in the experiment that you are evaluating and can manipulate; example: giving a specific amount of drug to a subset of the members of your sample prior to testing.

Intercept the value of y where a line crosses the vertical axis.

Interquartile range measurements lying between the upper quartile (75th percentile) and the lower quartile (25th percentile).

Lower quartile the lower 25th percentile of a group of measurements.

Measures of variation descriptive measurements of variance; standard deviation, standard error.

Middle quartile the 50th percentile of a group of measurements.

Mixed model design experimental design where some treatments are fixed effects and others are random effects.

Multivariate when there are two or more dependent variables being predicted by two or more independent variables.

Mutually exclusive events such that the occurrence of one precludes the occurrence of the other.

Negative relationship a relationship between two variables such that as one increases in value the other one decreases.

Parameter a specific characteristic about a population that one is trying to investigate; example: changes in cell firing rates, types of neurons in a given nucleus, total number of astrocytes, changes in mitochondrial respiration.

Population a complete or theoretical set of subjects or objects that are grouped according to some observable characteristic that one is trying to generalize to; it is often impossible to evaluate all of the members of a given population because it may be indefinite in size or extremely large that it is not feasible to evaluate every member; a population is often considered to be hypothetical; example: all of the young adult male Sprague–Dawley rats at the breeder.

Qualitative data data values that can be grouped into specific categories based on some defined characteristic that themselves have no real numerical value; example: species of rat in the vivarium.

Quantitative data data values that have a numeric value; example: the number of correct choices a rat makes in a maze.

Random effects model experimental design where it is assumed that the treatment levels represent a random sample from a population of treatment levels.

Region of acceptance the area of a probability curve where a computed test statistic will lead to acceptance of the null hypothesis.

Relative frequency the percentage of the total number of data points that are in any specific group; example: the relative frequency of the 47 male rats in the first room of the vivarium is 0.09 and the relative frequency of the 93 male rats in the second room of the vivarium is 0.18.

Sample a subset of the population that you are trying to generalize to; the goal is to use a sample to make a statistical inference concerning the hypothetical population; example: the 18 young adult male Sprague–Dawley rats that you purchased from the breeder.

Sampling distribution a theoretical probability distribution obtained with repeated sampling of the population.

Skewness refers to the shape of a distribution; a positively skewed distribution has most of the data part on the left side of the mean; a negatively skewed distribution has most of the data points on the right side of the mean; a symmetrical distribution has about equal numbers of data points on both sides of the mean.

Statistic some type of characteristic of a sample; example: the mean number of astrocytes in the red nucleus of an aged Sprague–Dawley rat; a given population or even a sample can have multiple statistics.

Statistical dispersion an indication of the variability within a set of data; standard deviation, variance.

Systematic error the consistent under- or overestimation of a true value because of poor sampling techniques.

Test statistic a statistic whose purpose is to prove a test of some statistical hypothesis.

Unbiased sample a sample that is drawn from a population using a randomization procedure to select the elements of the sample.

Variable a parameter that is of interest in an experiment; example: the firing rate of a neuron.

Variance a measurement of the amount of spread of data points in a sample or population; the standard deviation is a measure of the variance in a sample.

Upper quartile the 75th percentile of a group of measurements.

Within-group variability differences among subjects treated in an identical fashion.

HOW TO READ STATISTICAL EQUATIONS

Even though the statistical software does all the work for you, it is important just to be able to understand some of the basics about statistical equations.

$$\frac{1}{n}\sum_{r=1}^{R}\sum_{c=1}^{C}T_{rc}^2 - \frac{1}{nC}\sum_{r=1}^{R}T_{r.}^2 - \frac{1}{nR}\sum_{c=1}^{C}T_{.c}^2 + \frac{T^2}{N} \qquad (10.1)$$

They are really not as intimidating as they may look. Often, when you stop by to chat with your friendly biostatistician, he or she may jot down a couple of equations just to make a point. Also, you may look at a statistics book just to check something out and they always include equations. Below are some of the basics about what the different symbols mean.

One of the most basic of all statistical notations is:

$$\sum_{i=1}^{N}X_i \qquad (10.2)$$

This is the basic summation notation and simply means adding up a series of numbers. The numbers above and below the Greek symbol define the limits of the operation. The N on top of the Greek symbol simply indicates how many numbers are to be added up. If nothing is stated then it is assumed that all of the numbers are used. The $i = 1$ below the Greek symbol simply means the numbers are taken beginning with 1. For example let us say you have the numbers: **8, 12, 6, 15, 10, 11** and you want to sum them. Statistically it would be written:

$$\sum_{i=1}^{6}X_i \qquad (10.3)$$

You have six numbers and add them up from 1 to 6 and in this case it equals **62**. Sometimes a statistical text will omit the top and bottom numbers and simply write the equation as with it being understood that all the numbers will be added starting with 1:

$$\sum X_i \tag{10.4}$$

In other situations you will see a notation where the numbers to be added will be multiplied by a constant as indicated by the letter "*c*":

$$\sum_{i=1}^{N} cX_i \tag{10.5}$$

It is also very common to see a notation that indicates the sum of pairs of numbers that are multiplied by each other. Consider that you have scores from five rats that have maze scores on different days. Day 1 scores are 8, 5, 9, 4, 7 and scores for Day 2 are 4, 6, 7, 11, 5. This would be represented in statistical notation

$$\sum_{i=1}^{N} X_i Y_i \tag{10.6}$$

and means $8 \cdot 4 + 5 \cdot 6 + 9 \cdot 7 + 4 \cdot 11 + 7 \cdot 5 = 204$.

In some of the statistical equations, one number of a pair is squared before being multiplied by a second number and it is written.

$$\sum_{i=1}^{N} X_i^2 Y_i \tag{10.7}$$

Here is the statistical notation for the commonly used standard deviation:

$$S = \sqrt{\frac{\sum (X - \overline{X})^2}{N - 1}} \tag{10.8}$$

Usually if you know what some of the symbols stand for it is not that difficult to figure out what is happening in the statistical formula. Of course having some knowledge of basic algebra helps.

IMPORTANT STATISTICAL SYMBOLS

Symbol	What it signifies
μ	Population mean
c	A constant
C	Number of columns in a table
d	Difference between the mean of a subgroup and the mean of all groups $\overline{X}_i - \overline{X} = d_i$
D	Difference between paired measurements
$\overline{D}$	Mean difference between paired measurements
df, DF	Degrees of freedom df_B between-groups df, df_R row df, df_W within-group df, df_{RC} row X column df
F	Ratio of two sample variances
f	Frequency in a distribution
f_e	Expected frequency
f_o	Observed frequency
H_o	Null hypothesis
H_A	Alternative hypothesis
H_1	Alternative hypothesis
i, j, k	Subscripts used to identify particular observations in a group
k	Number of groups or means
K	Kendall's coefficient
N	Number of observations in a sample
n	Number of observations in a sample
n_k	Number of scores in the kth group
Np	Number of individuals in a population
O	Observed frequency in a calculation of chi-square (X^2)
P	P value; percentile point if P_{25} it is the 25th percentile
p	Probability, sample probability of two mutually exclusive classes
R	Number of rows in a table
r	Pearson product moment correlation coefficient; can denote rth row in a set of R rows
r^2	Coefficient of determination
SD	Standard deviation of a sample

cont'd

Symbol	What it signifies
s	Standard deviation of a sample
s_X	Standard deviation of the X variable
s_Y	Standard deviation of the Y variable
$s_{\overline{X}}$	Estimated standard error of the mean
$s_{\overline{X}_1 - \overline{X}_2}$	Estimated standard error of the difference between sample means
s^2	Variance estimate of a sample; square of the standard deviation
s_B^2	Between-groups estimate of variance
s_w^2	Within-groups estimate of variance
SE	Standard error
SS	Sum of squares of a sample
SEM	Standard error of the mean
SS_B	Between-groups sum of squares
SS_W	Within-groups sum of squares
T	True value of an observation or measurement; in an ANOVA it signifies the sum of observations (T_f is the sum of observations in the fth group)
t	Student's statistic
U	Statistic computed in the Mann–Whitney U test
X	Variable in its original unit of measurement; X_3 means the third data point; X_{2-4} means the fourth data point in the second set
X'	Predicted X value
$\overline{X}$	Arithmetic mean of a sample; the bar over the top of a letter always signifies a sample mean
α	Threshold for rejecting the null hypothesis; probability of a Type I error
β	Probability of a Type II error
μ	Population mean
ρ	Correlation coefficient of a population
σ	Standard deviation of a population
σ^2	Variance of a population
τ	Kedall's (tau) coefficient of a rank correlation
Σ	Sum of something
$a > b$	a is greater than b, e.g., a is greater than 8

Continued

cont'd

Symbol	What it signifies
$a < b$	a is less than b, e.g., a is less than 15
$a \geq b$	a is greater than or equal to b
$a \leq b$	a is less than or equal to b
$\neq$	Is not equal to 11 is not equal to 14
$=$	Equal to
$\lvert a \rvert$	Absolute value of a, e.g., absolute value of -7 $\lvert -7 \rvert = 7$
$\sqrt{a}$	Square root of a $\sqrt{144} = 12$
a^2	Square a $5^2 = 25$

Index

Printed in the United States
By Bookmasters